Fundamentals of Food Microbiology

Fundamentals of Food Microbiology

Conard Nash

Callisto Reference

www.callistoreference.com

Callisto Reference,
118-35 Queens Blvd., Suite 400,
Forest Hills, NY 11375, USA

Visit us on the World Wide Web at:
www.callistoreference.com

ISBN: 978-1-64116-240-1 (Hardback)

Printed in China.

Cataloging-in-Publication Data

Fundamentals of food microbiology / Conard Nash.
p. cm.
Includes bibliographical references and index.
ISBN 978-1-64116-240-1
1. Food--Microbiology. 2. Food spoilage. 3. Dairy microbiology.
I. Nash, Conard.

QR115 .F86 2019
664.001 579--dc23

Table of Contents

Preface

The study of microorganisms inhabiting, creating or contaminating food is called food microbiology. It also involves the study of food spoilage, food-borne diseases and the causative microorganisms and pathogens. Pathogens such as bacteria and viruses are easily transmitted through food, causing food contamination. However, many microbes like bacteriophages and probiotic bacteria and their products can be used for combating these pathogens. An important aspect of the study of food microbiology is the study of microbes for producing fermented foods and probiotics. Microbial biopolymers such as alginate and polyglutamic acid are used in the food industry as thickeners, while yeast is used to brew beer, make wine or leaven bread. Lactic acid bacteria play a crucial role in making cheese, yogurt, pickles, kimchi or fermented sausages. This book elucidates the concepts and innovative models around prospective developments with respect to food microbiology. It studies, analyzes and upholds the pillars of this discipline and its utmost significance in modern times. For all those who are interested in food microbiology, this textbook can prove to be an essential guide.

Given below is the chapter wise description of the book:

Chapter 1- Food microbiology is the science concerned with the study of microorganisms that create, inhabit, contaminate or spoil food. This is an introductory chapter, which will introduce briefly all the important microorganisms with respect to food microbiology, such as viruses, bacteria, fungi and protozoa, besides others.

Chapter 2- The edibility of food is reduced considerably with time due to spoilage caused by microbial growth. The chapter closely examines the causative factors of food spoilage especially due to microbial growth. It includes essential topics relevant to the understanding of this subject, such as growth factors of microorganisms, meat spoilage and wine fault.

Chapter 3- Fermentation is an important practice in food processing. It is the process of conversion of carbohydrates into alcohol or organic acids under anaerobic conditions with the aid of microorganisms, like yeasts and bacteria. The two primary processes of lactic acid fermentation and alcoholic fermentation have been covered in extensive detail in this chapter. It also elucidates about the different kinds of fermented foods such as dairy products, sauerkraut, ngari, fermented sausages, etc.

Chapter 4- Foodborne illness is a health condition that occurs due to the consumption of spoiled food. The major causes of food spoilage are pathogenic bacteria, parasites and viruses. This chapter discusses in detail about microbial foodborne diseases, infections and intoxications as well as the ways to prevent and treat them. All the diverse bacterial pathogens that cause food poisoning, such as Salmonella spp., Bacillus cereus, Campylobacter jejuni, Staphylococcus aureus, Escherichia coli, etc. have also been carefully analyzed in this chapter.

Indeed, my job was extremely crucial and challenging as I had to ensure that every chapter is informative and structured in a student-friendly manner. I am thankful for the support provided by my family and colleagues during the completion of this book.

Conard Nash

Chapter 1

Food Microbiology and Important Microorganisms

Food microbiology is the science concerned with the study of microorganisms that create, inhabit, contaminate or spoil food. This is an introductory chapter, which will introduce briefly all the important microorganisms with respect to food microbiology, such as viruses, bacteria, fungi and protozoa, besides others.

Food Microbiology

Microbiology is important to food safety, production, processing, preservation, and storage. Food microbiology use a wide variety of modern technologies from fields including immunology, microbiology, and molecular biology. Microbes such as yeasts, molds, and bacteria are being used for the production of foods and food ingredients. Beneficial microbes are exploited in the fermentative production, processing, and preservation of many foods and beverages. Spoilage microorganisms cost food producers, processors, and consumers millions of dollars annually in lost products. Lost productivity resulting from illness caused by foodborne microorganisms is an enormous economic burden throughout the world. The study of food microbiology includes understanding not only the factors influencing the growth of microorganisms in food systems but also the means of controlling them.

Microbiological analysis is important to determine the safety and quality of food. For many years, detection and identification of microorganisms in foods, animal feces, and environmental samples have relied on cultural techniques regarded as the "gold standard". Conventional methods are labor intensive, time consuming, and costly, and advances in these methods have been limited to the development of instruments such as the Stomacher or Pulsifier for sample processing, improved liquid and selective/differential agar media, instruments for plating and counting bacteria, and identification test kits. More recently, advances in biotechnology have led to the development of "rapid methods" that minimize manipulation, provide results in less time, and reduce cost. Rapid methods generally include immuno-based and DNA-based assays. Immunological or antibody-based assays include enzyme linked-immunosorbent assays (ELISA) and immunochromatographic or "dipstick" assays. Genetic methods include the polymerase chain reaction (PCR), DNA hybridization, and DNA microarrays, also known as GeneChips.

Beneficial Role of Microorganisms in Food Industry

Microorganisms play an important role in food industry. They are used in production of various food products, and are also responsible for food spoilage thereby causing intoxication and diseases.

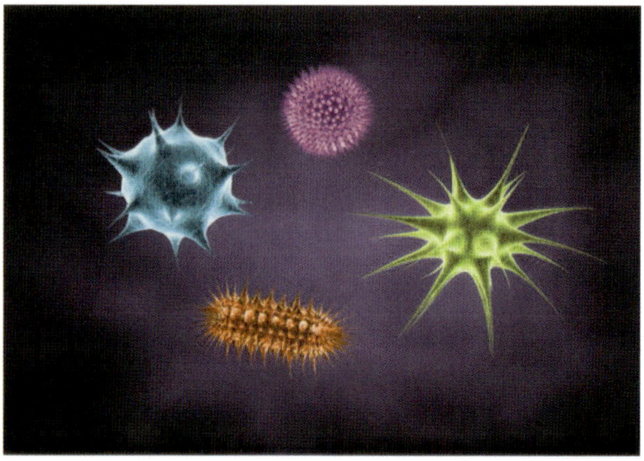

Microorganisms

Microbial contamination of food products takes places usually on the way from the field to the processing plant, or during processing, storage, transport and distribution or before consumption. The microorganisms that cause food spoilage and also find the maximum exploitation in production of food and food products are mainly bacteria, molds and yeasts.

Bacteria

Bacteria are the largest group of unicellular microorganisms. The shapes of medically important bacteria are classified into-cocci, or spherical cells; bacilli, or cylindrical or rod shaped cells; and spiral or curved forms. The pathogenic or disease causing bacteria are usually gram negative, however, three gram-positive rods are known to cause food intoxications : *Clostridium botulinum, C. perfringens, and Bacillus cereus.*

Some of the other most common bacteria causing food spoilage, infections and disease are *Acinetobacter, Aeromonas, Escherichia, Proteus, Alcaligenes, Flavobacterium, Pseudomonas, Arcobacter, Salmonella, Lactococcus, Serratia, Campylobacter, Shigella, Citrobacter, Listeria, Staphylococcus, Micrococcus, Corynebacterium, Vibrio Enterobacter, Paenibacillus, Weissella, Enterococcus, Yersinia.*

Different strains of bacteria are also used in production of various food and dairy products. Strains of Streptococcus, *Lactobacillus Bifidobacterium*, Erwiniaetc. are used in the production of fermented food and dairy products. *Streptococcus thermophilus* and *Lactobacillusbulgaricus* are used to produce yogurt.

Molds

Molds are multicellular filamentous fungi whose growth on foods is usually readily recognized by their fuzzy or cottony appearance. They are mainly responsible for food spoilage at room temperature 25- 30oC and low pH, and have minimum moisture requirement. Molds can rapidly grow on grains and corns when these products are stored under moist conditions. Molds require free oxygen for growth and hence grow on the surface of contaminated food. Molds also find their use in manufacturing of different foods and food products. They are used in ripening of various types of food products as cheese (e.g. Roquefort,Camembert). Molds are also grown as feed and food and are employed to produce ingredients such as enzymes like amylase used in making bread or citric acid used in soft drinks. Molds are major contributors in the ripening of many oriental foods. A species of *Bothrytiscinerea*, is used in rotting of grape for production of wine. Lactic fermentations using molds results in a unique Finnish fermented milk called viili.

Yeasts

Yeasts have the ability to ferment sugars to ethanol and carbon-dioxide and hence they are extensively in food industry. The most commonly used yeast, the baker's yeast is grown industrially. *Saccharomyces carlsbergensis* is most commonly used in fermentation of most beers. The other yeast strains of importance are

Brettanomyces, Schizosaccharomyce, Candida, Cryptococcus, Debaryomyces, Zygosaccharomyces, Hanseniaspora, Saccharomyces.

Food safety is a matter that affects anyone who eats food. Whether or not a person consciously thinks about food safety before eating a meal, a host of other people have thought about the safety of that food, from farmers to scientists to company presidents to federal government officials and public health officials. Ensuring the safety of food is a shared responsibility among producers, industry, government, and consumers. Safe food is food that is free not only from toxins, pesticides, and chemical and physical contaminants, but also from microbiological pathogens such as bacteria, parasites, and viruses that can cause illness.

Those working in the field of food safety are most concerned about microbial foodborne illness, a widespread but often unrecognized sickness that affects most people at one time or another. At least four factors are necessary for foodborne illness to occur: (1) a pathogen; (2) a food vehicle; (3) conditions that allow the pathogen to survive, reproduce, or produce a toxin; and (4) a susceptible person who ingests enough of the pathogen or its toxin to cause illness. The symptoms often are similar to those associated with the flu—nausea, vomiting, diarrhea, abdominal pain, fever, headache. Most people have experienced foodborne illness, even though they might not recognize it as such, instead blaming it on the stomach flu or a twenty-four-hour bug. Usually symptoms disappear within a few days, but in some cases there can be more long-lasting effects such as joint inflammation or kidney failure. In the most severe cases people die from foodborne illness.

Current estimates of foodborne illness in the United States are 76 million cases, 325,000 hospitalizations, and 5,194 deaths from foodborne pathogens per year. In cases when the pathogen is identified, bacteria cause 30 percent of foodborne illnesses, parasites 3 percent, and viruses 67 percent. But as far as deaths are concerned, bacterial pathogens are the leading cause of death, with 72 percent of total foodborne illness deaths attributable to bacteria. Fatality rates for two bacteria are particularly high; for *Listeria* 20 percent of the people may die, and for *Vibrio vulnificus* 39 percent. Just six pathogens account for over 90 percent of the deaths associated with foodborne illness: *Salmonella* (31 percent), *Listeria* (28 percent), Toxoplasma (21 percent), Norwalk-like viruses (7 percent), *Campylobacter* (5 percent), and *Escherichia coli* (3 percent). According to FoodNet data from 1996–1997, each person in the United States suffers 1.4 episodes of diarrhea per year. With a U.S. population of 267.7 million persons, that works out to 375 million episodes per year, many of them related to eating unsafe food. Factors that contributed the most to foodborne illness are improper holding temperatures, inadequate cooking, contaminated equipment, food from unsafe sources, and poor personal hygiene.

Factors Influencing the Safety of Food

Stories of foodborne illness have become much more prevalent throughout the world. Is food less safe than it used to be, and if so, what factors account for this? News travels fast these days, both electronically and through the news media. What were once isolated events and stories, now reach millions within hours. Diagnostic techniques are constantly improving, allowing for identification of diseases, foodborne and otherwise, that would have been of unknown origin in the past. But even considering these facts, public-health officials believe that the risk of foodborne illness has increased over the past twenty years (GAO, 1996). Some threats to food safety have been around since ancient times, while others are newer, the result of changing demographics and lifestyles, production practices, and even evolution of microorganisms themselves.

- Microbial Biopolymers

Several microbially produced biopolymers are used in the food industry.

- Alginate

Alginates can be used as thickening agents. Although listed here under the category 'Microbial polysaccharides', commercial alginates are currently only produced by extraction from brown seaweeds such as *Laminaria hyperborea* or *L. japonica*.

- Poly-γ-glutamic Acid

Poly-γ-glutamic acid (γ-PGA) produced by various strains of *Bacillus* has potential applications as a thickener in the food industry.

Yeast

Yeast are single-celled fungi. As fungi, they are related to the other fungi that people are more familiar with, including: edible mushrooms available at the supermarket, common baker's yeast used to leaven bread, molds that ripen blue cheese, and the molds that produce antibiotics for medical and veterinary use.

Yeast cells are egg-shaped and can only be seen with a microscope. It takes 20,000,000,000 (twenty billion) yeast cells to weigh one gram, or 1/28 of an ounce, of cake yeast.

A Tiny Organism with a Long Name

The scientific name for the yeast that baker's use is Saccharomyces Cerevisiae, or "sugar-eating fungus". A very long name for such a tiny organism. This species of yeast is very strong and capable of fermentation, the process that causes bread dough to rise.

Working of Yeast

Through the process of fermentation, yeast converts sugars into carbon dioxide and alcohol. These two byproducts make yeast an extremely useful tool in food production.

Carbon dioxide is what gives alcoholic beverages such as beer and champagne their characteristic bubbles and is also responsible for rising bread. As yeast begins to metabolize sugars in bread dough, the carbon dioxide gas is trapped within the gluten strands, creating bubbles, and causing a leavening action. In beverages, the carbon dioxide is trapped within the liquid by the pressure of the sealed container. When the container is opened, the pressure is released and the carbon dioxide begins to release or bubble.

Alcohol, the other byproduct of yeast fermentation, is also produced during the bread making process but evaporates as the bread bakes. When making alcoholic beverages, yeast is allowed to ferment for a much longer period of time, allowing it to produce more alcohol.

Culinary uses for Yeast

- Beer: Yeast is added to malted grains and allowed to ferment in order to produce alcohol. The type of yeast used will affect the type and flavor of beer produced. S. cerevisiae, also known as a "top fermenting" or "top cropping" yeast, ferments at a higher temperature and produces sweet or fruity beers. Bottom croppers, such as Saccharomyces pastorianus, ferment at lower temperatures and are used to make lagers.

- Wine: Yeast is naturally present on the skins of grapes and can be used to naturally ferment grape juice into wine. Despite the naturally present yeast, most wines today have pure culture (usually S. cerevisiae) added to them to produce a more consistent and controllable result. There are many different strains of S. cerevisiae and each will produce a unique flavor characteristic in a finished wine.
- Bread: Records of using yeast as a leavening agent date back to the ancient Egyptians, though the form of yeast used has changed over time. Many varieties are available for use in bread making, such as fresh yeast cakes, bakers yeast, active dry, instant, or yeast starter.
- Nutrition: Raw yeast is not recommended for consumption in quantity as it can proliferate wit-hin your body, but once pasteurized, yeast offers a host of nutritional benefits. Yeast is a good source of protein, B vitamins, and minerals. Yeast intended as a nutritional supplement is often sold as "nutritional yeast" or "yeast extract." Yeast extracts, such as Marmite, have a cult following due to their unique flavor and variety of uses.

Beyond Brewing, Baking and Wine Yeasts

Although research on the contribution of S. cerevisiae to beer, bread and wine fermentations continues to be a focus, there is expanding interest in the role of yeasts in other products.

It is now well recognized that yeasts make important contributions to the process of cheese maturation, where various strains of D. hansenii, Yarrowia lipolytica, Kluyveromyces marxianus and S. cerevisiae frequently grow to high populations. They contribute to the development of cheese flavour and texture through proteolysis, lipolysis, utilization of lactic acid, fermentation of lactose and autolysis of their biomass. In a similar way, D. hansenii, Y. lipolytica and various Candida species affect flavour, texture and colour development in fermented salami style sausages and country cured hams. Many breads, especially sour dough varieties, are still produced by traditional fermentation processes where no commercial strains of baker's yeast are added. Although indigenous strains of S. cerevisiae are prominent in many of these fermentations, other yeasts are significant and include Saccharomyces exiguus, Candida milleri, Candida humilis, Candida krusei (Issatchenkia orientalis), Pichia anomala, Pichia membranifaciens and Y. lipolyitica. These yeasts grow in cooperation with lactic acid bacteria, giving distinctive flavours to the final product.

High-value cash crops such as cocoa beans and coffee beans also undergo processes that involve the action of yeasts. Coca beans must be fermented to generate the precursors of chocolate flavour, and various species of Saccharomyces, Hanseniaspora, Candida, Issatchenkia and Pichia contribute to the process. Coffee beans are processed to remove pulp and other mucilaginous materials that surround the seeds, and species of Candida, Saccharomyces, Kluyveromyces, Saccharomycopsis, Hanseniaspora, Pichia and Arxula have been associated with these fermentations. A vast array of traditional fermented foods and beverages are produced in African, Asian and South American countries from raw materials such as maize, wheat, cassava, rice, soy beans and fruit. Fermentation is essential in contributing to the quality, safety and nutritional value of these products. Aspects of their microbial ecology are just starting to emerge, and demonstrate important contributions from numerous yeast species.

Collectively, the ecological studies of yeasts in products other than beer, bread and wine are providing the knowledge base for developing a new generation of yeast starter cultures, beyond S. cerevisiae.

Microbial Interactions and Biocontrol

Yeasts rarely occur in food and beverage ecosystems as single cultures. Exceptions occur in highly processed products where spoilage outbreaks by single, welladapted species are known: for example, Z. rouxii in high sugar products.

Generally, most habitats are comprised of a mixture of yeasts, bacteria, filamentous fungi and their viruses, and product quality is determined by the interactive growth and metabolic activity of the total microflora. Even within yeasts themselves, there can be significant species and strain interactions that impact on the population dynamics of the ecosystem. The diversity and complexity of these microbial interactions is just beginning to emerge.

A network of yeast–yeast interactions occurs in most ecosystems, and is observed in fermentations of wine, cheese, meat, and cocoa beans. These interactions manifest themselves as the successive growth and death of different yeast species and strains within each species, as the fermentation progresses. The mechanisms underlying these ecological shifts are numerous. Explanations include the different rates of nutrient transport and uptake by the different species and strains, their sensitivities to metabolic end products (e.g. ethanol), and responses to killer toxins. Cell–cell interactions might also occur through the production of quorum sensing molecules and unexplained spatial phenomena. Defining the metabolic outcomes of these interactions and their impact on product quality remains a greater challenge, as demonstrated by the interactive effects of S. cerevisiae and Saccharomyces bayanus strains on the chemical composition and flavour of wines.

Interactions between yeast and bacteria are often seen as the inhibitory effects of yeasts on bacteria through ethanol production; however, the relationships are much broader than this. The death and autolysis of yeast cells releases vitamins and other nutrients that stimulate the growth of important flavour-enhancing bacteria, such as the malolactic bacteria in wine fermentations, staphylolcocci, micrococci and brevibacteria during cheese maturation, and lactic acid bacteria during sour dough fermentations. Ethanol, produced by yeasts during cocoa bean fermentations, stimulates the growth of acetic acid bacteria that oxidize the ethanol to acetic acid. This acid is essential for killing the cocoa beans (seeds) and triggering endogenous bean metabolism that generates the precursors of chocolate flavour. Some yeasts utilize the organic acids that occur in cheeses, fruit products and salad dressings, causing an increase in product pH and growth of spoilage and pathogenic bacteria. Some bacteria are antagonistic towards yeasts. Excessive growth of lactic acid bacteria and acetic acid bacteria on grapes produces acetic acid and other substances that inhibit the growth of yeasts in grape juice, causing stuck or sluggish wine fermentations and loss of process efficiency.

Interactions between yeast and fungi have not been widely studied, except in the context of biocontrol. Fungal growth on wine grapes produces substances that inhibit the growth of yeasts during grape juice fermentation. By contrast, some yeasts improve the growth of Penicillium spp. during the maturation of cheeses. Several species within the genera Candida, Pichia, Metschnikowia, Cryptococcus and Pseudozyma have strong antifungal properties mediated through the production of lytic enzymes, toxic proteins, toxic fatty acids and ethyl acetate, and have potential for the biocontrol of fungi. Commercial preparations of some species are now available for the pre- and post-harvest control of fruit, vegetable and grain spoilage fungi.

Yeasts and Food Safety

As part of daily life, humans consume large populations of yeasts without adverse impact on their health. Unlike bacteria and viruses, yeasts are rarely associated with outbreaks of foodborne gastroenteritis, intoxications or other infections.

Significant 'lay' literature connects the dietary intake of yeasts with a range of gastrointestinal, respiratory, skin, migraine and even psychiatric disorders. Overgrowth of yeasts in the gastrointestinal tract might contribute to the development of these disorders, but immune reactions to yeast cell wall polysaccharides and responses to yeast-produced amines and sulfur dioxide could also occur. The connection between yeast, the human response and food is largely based on dietary observations. If foods suspected to contain yeasts or their products are removed from the diet, the adverse responses disappear, but return when such foods are reintroduced.

Yeasts are not aggressive, infectious organisms, but some species such as Candida albicans and Cryptococcus neoformans are opportunistic pathogens that cause a range of mucocutaneous, cutaneous, respiratory, central nervous system and organ infections, as well as general fungemia. Individuals with weakened health and immune systems are at greatest risk, and include cancer, AIDS and hospitalized patients, and those undergoing treatment with immunosuppressive drugs, broad spectrum antibiotics and radio- chemotherapies. The greater frequency of such individuals in thecommunity has led to increased reporting of yeast infections. Moreover, an increasing number of yeast species has been implicated, includingmany found in foods (e.g. S. cerevisiae, C. krusei, C. famata, P. anomola, Rhodotorula spp. Infections caused by S. cerevisiae are notable because of its extensive use in the food industry, and infections with this yeast have been reported in immunocompetent individuals. It is thought that hospitalized patients become exposed to high levels of yeasts through the biofilms they form on catheters and other invasive devices, and that these yeasts probably originate from the hands of hospital workers and the foods brought into the hospital environment. More research is needed to establish stronger linkages between the role of foods in contributing to yeast infections. Information is needed on the survival and growth of yeasts throughout the gastrointestinal system, the potential for yeasts to translocate from the gastrointestinal tract to the blood system, and the general occurrence of yeasts 'in the hospital and health care environments. The circumstances whereby a non-pathogenic yeast, such as S. cerevisiae, becomes pathogenic also require investigation.

Probiotic and other Health Benefits

Probiotics are viable microorganisms that are beneficial to consumers when ingested in appropriate quantities. Although certain species of lactic acid bacteria are prominent as probiotic organisms, there is increasing interest in yeasts as probiotics. S. cerevisiae var boulardii has been used for many years as an oral biotherapeutic agent for treating a range of diarrheal disorders. This species colonizes the intestinal tract where, in a probiotic function, it combats diarrhoea-causing bacteria. Food carrier systems for this yeast need to be developed for its commercial application as a probiotic, but technical obstacles have been encountered. When incorporated into some products, it caused gassy, ethanolic spoilage and off-flavours. Of greater concern, are reports of fungemia infections caused by S. boulardii. Other yeasts mentioned as potential probiotics include D. hansenii, Kluy. marxianus, Y. lipolytica, I. orientalis, P. farinosa and P. anomala, but further research is required. Yeasts are increasingly used as probiotics in the livestock and aquaculture industries.

Yeast products, principally derived from S. cerevisiae, have been used for many years as ingredients and additives in food processing. These products include flavourants, enzymes, antioxidants, vitamins, colourants and polysaccharides. Three points are worthy of mention. First, many of these products are prepared from yeast cells after they have been processed by autolysis. Despite its commercial significance, molecular understanding of yeast autolysis is still very limited and more research is needed to optimize this process. Second, most products are derived from S. cerevisiae. The yield and range of products could be increased by screening for their presence in other yeast species and strains, as demonstrated for the vitamin folic acid, cell wall polysaccharides and autolysates. Finally, there remains undiscovered bioactivity and functionality in yeast products. Whereas the glucan polysaccharides from the walls of S. cerevisiae were originally valued for their water-binding and rheological functionalities, it is now recognized that they can stimulate the immune system, lower serum cholesterol, exhibit antitumour activity, and adsorb substances such as mycotoxins.

Mold

Molds are microscopic fungi that live on plant or animal matter. No one knows how many species of fungi exist, but estimates range from tens of thousands to perhaps 300,000 or more. Most are filamentous (threadlike) organisms and the production of spores is characteristic of fungi in general. These spores can be transported by air, water, or insects.

Unlike bacteria that are one-celled, molds are made of many cells and can sometimes be seen with the naked eye. Under a microscope, they look like skinny mushrooms. In many molds, the body consists of:

- root threads that invade the food it lives on,
- a stalk rising above the food, and
- spores that form at the ends of the stalks.

The spores give mold the color you see. When airborne, the spores spread the mold from place to place like dandelion seeds blowing across a meadow.

Molds have branches and roots that are like very thin threads. The roots may be difficult to see when the mold is growing on food and may be very deep in the food. Foods that are moldy may also have invisible bacteria growing along with the mold.

Here are common types of mold found in homes and businesses:

Absidia	*Curvularia*	*Gliomastix*	*Phoma*
Acremonium	*Emericella*	*Memnoniella*	*Scopulariopsis*
Alternaria	*Epicoccum*	*Mucor*	*Sistotrema*
Aspergillus	*Eurotium*	*Myrothecium*	*Stachybotrys*
Aureobasidium	*Fusarium*	*Oidiodendron*	*Trichoderma*
Chaetomium	*Geomyces*	*Paecilomyces*	*Ulocladium*
Chrysonilia	*Geotrichum*	*Penicillium*	*Wallemia*
Cladosporium	*Gliocladium*	*Phialophora*	

Conditions for Mold Growth

Mold needs four things in order to grow: water, food, suitable air quality and temperature. Food that contains any kind of water or fluid is susceptible to mold growth. In addition, mold can only grow if it has food readily available to feed itself and grow. Mold is a fungus that feeds off of dead or dying organic matter, and can be destructive to your health and food quality. Mold will grow best in damp, dark and cool conditions, but can also grow in warmer temperatures as well. Mold grows best between 55 to 70 degrees Celsius.

Ways in which Mold Grows on Food

Tiny mold spores are all around us in the air, which is not harmful to our health in moderation. Once a spore lands on a surface, it searches for water and nutrients to feed off of. Food is able to grow mold easily because it is often kept in the perfect environment to foster mold growth. The temperature is usually about right, air quality is good, and the food itself provides the nutrients and water the mold needs in order to grow. As the spore takes root, it begins to spread and create more spores and spread quickly on the surface of your food. Some molds can take over your food in a matter of 12 to 24 hours, while others may take weeks.

Dangers of Mold on Food

Mold can be very dangerous if eaten. Mold is perhaps most common on bread, cheese, and fruits and vegetables left out in the kitchen. Eating mold on any item has the potential to make you very sick. Simply cutting off the moldy part does not render the food safe to eat. Mold has the ability to penetrate deep into the food and not just fester on the surface. If you do spot mold on your food, it is best to inspect the entire serving and not just one spot. In addition, if the food you are eating is part of a package of multiple servings, you should check all of them – mold can spread very quickly and infest an entire package of food.

Molds produce irritating substances that may act as allergy causing substances (allergens) in sensitive individuals. Furthermore, some molds produce toxic substances known as mycotoxins, but mold itself is not poisonous or toxic. The term "toxic mold" therefore, refers to the fact that certain kinds of mold can produce mycotoxins. The conditions under which some molds produce toxins is not understood, and the presence of mold, even a mold that is capable of producing toxins, does not always imply that toxins are being produced or that a health risk or problem is present. Mold may not cause any health problems, or it may lead to allergy or other symptoms in people, including adults and children, who are sensitive to molds.

Allergic reactions to mold are the most common health effects of mold and are therefore the greatest health risk related to mold. Allergic reactions may happen immediately or develop after a period following exposure. Both growing mold and mold spores may lead to allergic reactions. Symptoms and signs of mold allergy may include

- sneezing,
- runny nose,
- coughing,
- wheezing,
- watery eyes,
- redness of the eyes,
- itchy eyes,
- skin irritation, or rash.

Mold or mold spores may cause asthma attacks in people who have asthma and are allergic to mold. Even in some nonallergic individuals, mold can cause symptoms of irritation in the eyes, skin, and airways. For example, the "black mold" *Stachybotrys*, along with some other types of mold, produces toxins known as mycotoxins that can cause irritation of the skin and airways in susceptible individuals.

Sometimes, people may develop severe reactions to mold exposure. Symptoms of severe reactions, which are uncommon, include fever and difficulty breathing. People with compromised immune systems or patients with chronic lung disease can develop serious infections of the lungs due to molds.

It is not possible to predict the degree of severity of the health risks associated with mold in the home. Allergic individuals vary in their degree of susceptibility to mold, and any symptoms and health risks also depend upon the extent and exact type of mold that is present.

In 2004 guidelines update in 2009, the Institute of Medicine (IOM) reported there was sufficient evidence to link indoor exposure to mold with upper respiratory tract symptoms, cough, and wheezing in otherwise healthy people, although the report stated that was no evidence that mold causes medical conditions like asthma, bronchitis, or other respiratory conditions. Mold also worsens asthma symptoms in people who have asthma. Mold was also reported to be linked to hypersensitivity pneumonitis in individuals susceptible to this immunologic condition. This uncommon disease is similar to pneumonia and can develop in susceptible individuals after brief or prolonged exposure to mold, but there has been no conclusive evidence to prove this relationship.

Fungus

Fungus, plural fungi, any of about 99,000 known species of organisms of the kingdom Fungi, which includes the yeasts, rusts, smuts, mildews, molds, and mushrooms.

There are also many funguslike organisms, including slime molds and oomycetes (water molds), that do not belong to kingdom Fungi but are often called fungi. Many of these funguslike organisms are included in the kingdom Chromista. Fungi are among the most widely distributed organisms on Earth and are of great environmental and medical importance. Many fungi are free-living in soil or water; others form parasitic or symbiotic relationships with plants or animals.

Morphology of Fungi

(a) General. Fungi vary widely in size and shape, from unicellular, microscopic organisms to multicellular forms easily seen with the naked eye. Individual cells range from 1 μ to 30 μ. Microscopic fungi exist as either molds or yeasts or both. Internally, fungal cells are fairly typical eucaryotic cells.

(b) Molds. The molds form large multicellular aggregates of long branching filaments, called hyphae. There are vegetative hyphae and reproductive hyphae. Spores are borne on the reproductive hyphae. (Fungal spores should not be confused with bacterial spores that are resistant bodies formed for bacterial survival rather than reproductive purposes.) Spore size, shape and structure are used in the classification and identification of fungi. The tube-like hyphae are responsible for the fluffy appearance of the macroscopic mold colony. The hyphae and other structures combine to form an elaborate network called a mycelium.

(c) Yeasts. These are large (5 to 8 μ), single-celled organisms that rarely form filaments. Most yeasts reproduce by the asexual process of budding. Yeast colonies are usually characterized by a smooth surface similar to that of many bacteria.

Physiology of Fungi

(a) Nutrition. Most fungi contain complex enzymes and other chemical substances which, when diffused into the host, break down the complex substances available – wood, vegetation, leather, bread, and so forth – into simpler substances that can be used for food. The chemical products of digestion are, therefore, completed outside of the organism, and the fungus absorbs the end products.

(b) Reproduction. Fungi reproduce sexually or asexually, or both, depending upon the species and the environmental conditions. As the name implies, sexual reproduction is the result of the union of two spores. Most fungi reproduce both sexually and asexually. Those that produce only asexual spores are known as Deuteromycetes Fungi imperfecti. This group is important because it contains most of the pathogenic fungi. The yeasts reproduce both by spores and by a process known as budding, which is similar to binary fission. The yeast cell forms a small knoblike protrusion, or bud, that separates from the mother cell and grows until it reaches full size, at which time the process is repeated.

(c) Growth. Fungi grow well under the same conditions that favor the growth of bacteria – warmth and moisture. It is for this reason that fungal infections pose a serious problem to troops in the tropics. As the temperature decreases, fungal activity also decreases; however, the spores are very resistant to cold, some surviving freezing temperatures for long periods of time. On the other hand, fungi are easily killed at high temperatures.

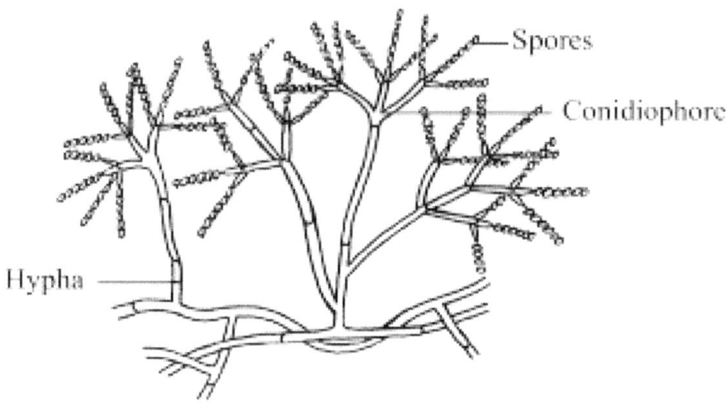

Typical mycelium of a fungus.

Classification of Fungi

Fungi are usually classified according to biological taxonomy based upon the type of hypha, spore, and reproduction. There are four classes of fungi:

(a) Class Phycomycetes. The algal fungi: bread molds and leaf molds. The only known mycosis (fungal disease) caused by fungi of this class is mucormycosis, a very rare fungal growth of the upper respiratory tract, bronchial mucosa, and lungs. It occurs largely as a complication of a chronic, debilitating disease, such as uncontrolled diabetes.

(b) Class Ascomycetes. The sac fungi: yeasts, mildews, and cheese molds. Fungi of this class are implicated in only three fungus diseases, all of which are rare.

(c) Class Basidiomycetes. Mushrooms, toadstools, rusts, and smuts. The only pathogens in this class are the mushrooms of the genus Amanita, which cause severe systemic poisoning (sometimes death) when eaten.

(d) Class Deuteromyceters. Fungi imperfecti: a heterogeneous collection of fungi without sexual reproduction. Most of the pathogens encountered in medical mycology belong to this class.

Ecology of Fungi

Relatively little is known of the effects of the environment on the distribution of fungi that utilize dead organic material as food (i.e., saprobic fungi). The availability of organic food is certainly one of the factors controlling such distribution. A great number of fungi appear able to utilize most types of organic materials, such as lignin, cellulose, or other polysaccharides, which have been added to soils or waters by dead vegetation. Most saprotrophic fungi are widely distributed throughout the world, only requiring that their habitats have sufficient organic content to support their growth. However, some saprotrophs are strictly tropical and others are strictly temperate-zone forms; fungi with specific nutritional requirements are even further localized.

Fungi are found in areas that have sufficient organic material and moisture to support their growth. For example, members of the genus Armillaria are often found in forests living on trees such as hardwoods or conifers

Moisture and temperature are two additional ecological factors that are important in determining the distribution of fungi. Laboratory studies have shown that many, perhaps the majority, of fungi are mesophilic, meaning they have an optimum growth temperature of 20–30 °C (68–86 °F). Thermophilic species are able to grow at 50 °C (122 °F) or higher but are unable to grow below 30 °C. Although the optimum temperature for growth of most fungi lies at or above 20 °C, a large number of species are able to grow close to or below 0 °C (32 °F). The so-called snow molds and the fungi that cause spoilage of refrigerated foods are examples of this group. Obviously, temperature relationships influence the distribution of various species. Certain other effects of temperature are also important factors in determining the habitats of fungi. Many coprophilous (dung-inhabiting) fungi, such as Pilobolus, although able to grow at a temperature of 20–30 °C, require a short period at 60 °C (140 °F) for their spores to germinate.

Mycology Definition

Mycology is the study of fungi, their relationships to each other and other organisms, and the unique biochemistry which sets them apart from other groups. Fungi are eukaryotic organism which belong to their own kingdom. Until advances in DNA technology, it was assumed that fungi were an offshoot of the plant kingdom. DNA and biochemical analysis has revealed that fungi are a separate lineage of eukaryotes, distinguished by their unique cell wall made of chitin and glucans which often surrounds multinucleated cells. Mycology is a necessary branch of biology because fungi is considerably different from both plants and animals.

Generally, many cases of poisonings arise from people eating wild, poisonous fungi. Such happenings occur as wrong identifications of mushroom species are made, and people unknowingly consume the toxic fungi. There are numerous edible mushrooms which have toxic lookalikes. This results in people assuming the species of mushroom that they have picked is harmless and safe to eat, when they are in reality poisonous.

It has been found that a high percentage of people poisoned by mushrooms are immigrants. Often, they mistake the toxic mushroom to be a safe, edible one found in their native country. Thus, the importance of educating people on lethal fungi is absolutely vital.

Hence, it is advisable for people to consume only mushrooms after consulting an expert. They should also thoroughly search their pickings to ensure that only edible mushrooms are in the collection. Although seemingly good sense, it is prudent to remind mushroom collectors to separate their edible pickings from their non-edible ones. Again, people have to be told to always cook their collections as some wild fungi are poisonous when consumed raw. To err on the side of caution, one should retain an uncooked specimen, so that in case of poisoning, medical personnel are able to identify the medication needed. This is as different species contain varying toxins.

Bacteria

Bacteria, singular bacterium, any of a group of microscopic single-celled organisms that live in enormous numbers in almost every environment on Earth, from deep-sea vents to deep below Earth's surface to the digestive tracts of humans.

Bacteria lack a membrane-bound nucleus and other internal structures and are therefore ranked among the unicellular life-forms called prokaryotes. Prokaryotes are the dominant living creatures on Earth, having been present for perhaps three-quarters of Earth history and having adapted to almost all available ecological habitats. As a group, they display exceedingly diverse metabolic capabilities and can use almost any organic compound, and some inorganic compounds, as a food source. Some bacteria can cause diseases in humans, animals, or plants, but most are harmless and are beneficial ecological agents whose metabolic activities sustain higher life-forms. Other bacteria are symbionts of plants and invertebrates, where they carry out important functions for the host, such as nitrogen fixation and cellulose degradation. Without prokaryotes, soil would not be fertile, and dead organic material would decay much more slowly. Some bacteria are widely used in the preparation of foods, chemicals, and antibiotics. Studies of the relationships between different groups of bacteria continue to yield new insights into the origin of life on Earth and mechanisms of evolution.

The Bacterial Cell

Bacteria as Prokaryotes

All living organisms on Earth are made up of one of two basic types of cells: eukaryotic cells, in which the genetic material is enclosed within a nuclear membrane, or prokaryotic cells, in which the genetic material is not separated from the rest of the cell. Traditionally, all prokaryotic cells were called bacteria and were classified in the prokaryotic kingdom Monera. However, their classification as Monera, equivalent in taxonomy to the other kingdoms—Plantae, Animalia, Fungi, and Protista—understated the remarkable genetic and metabolic diversity exhibited by prokaryotic cells relative to eukaryotic cells. In the late 1970s American microbiologist Carl Woese pioneered a major change in classification by placing all organisms into three domains—Eukarya, Bacteria (originally called Eubacteria), and Archaea (originally called Archaebacteria)—to reflect the three ancient lines of evolution. The prokaryotic organisms that were formerly known as bacteria were then divided into two of these domains, Bacteria and Archaea. Bacteria and Archaea are superficially similar; for example, they do not have intracellular organelles, and they have circular DNA. However, they are fundamentally distinct, and their separation is based on the genetic evidence for their ancient and separate evolutionary lineages, as well as fundamental differences in their chemistry and physiology. Members of these two prokaryotic domains are as different from one another as they are from eukaryotic cells.

Bacterial cells differ from animal cells and plant cells in several ways. One fundamental difference is that bacterial cells lack intracellular organelles, such as mitochondria, chloroplasts, and a nucleus, which are present in both animal cells and plant cells

Prokaryotic cells (i.e., Bacteria and Archaea) are fundamentally different from the eukaryotic cells that constitute other forms of life. Prokaryotic cells are defined by a much simpler design than is found in eukaryotic cells. The most-apparent simplification is the lack of intracellular organelles, which are features characteristic of eukaryotic cells. Organelles are discrete membrane-enclosed structures that are contained in the cytoplasm and include the nucleus, where genetic information is retained, copied, and expressed; the mitochondria and chloroplasts, where chemical or light energy is converted into metabolic energy; the lysosome, where ingested proteins are digested and other nutrients are made available; and the endoplasmic reticulum and the Golgi apparatus, where the proteins that are synthesized by and released from the cell are assembled, modified, and exported. All of the activities performed by organelles also take place in bacteria, but they are not carried out by specialized structures. In addition, prokaryotic cells are usually much smaller than eukaryotic cells. The small size, simple design, and broad metabolic capabilities of bacteria allow them to grow and divide very rapidly and to inhabit and flourish in almost any environment.

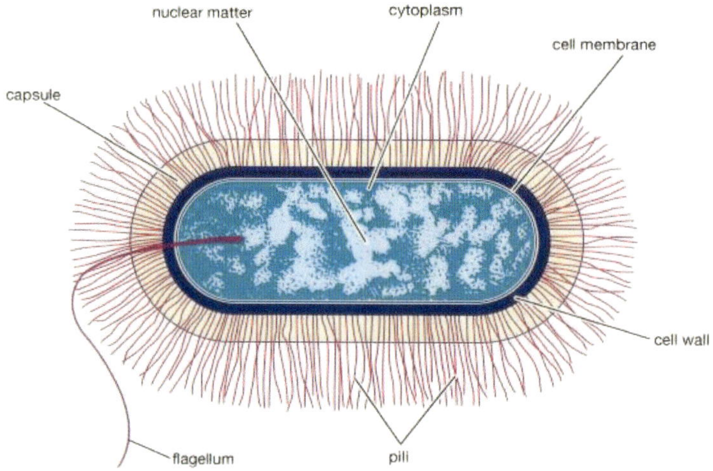

Schematic drawing of the structure of a typical bacterial cell of the bacillus type.Encyclopædia Britannica, Inc.

Prokaryotic and eukaryotic cells differ in many other ways, including lipid composition, structure of key metabolic enzymes, responses to antibiotics and toxins, and the mechanism of expression of genetic information. Eukaryotic organisms contain multiple linear chromosomes with genes that are much larger than they need to be to encode the synthesis of proteins. Substantial portions of the ribonucleic acid (RNA) copy of the genetic information (deoxyribonucleic acid, or DNA) are discarded, and the remaining messenger RNA (mRNA) is substantially modified before it is translated into protein. In contrast, bacteria have one circular chromosome that contains all of their genetic information, and their mRNAs are exact copies of their genes and are not modified

Classification of Bacteria

Before the invention of the DNA sequencing technique, bacteria were mainly classified based on their shapes - also known as morphology, biochemistry and staining

- i.e. either Gram positive or Gram negative staining. Nowadays, along with the morphology, DNA sequencing is also used in order to classify bacteria. DNA sequencing helps in understanding the relationship between two types of bacteria i.e. if they are related to each other despite their different shapes. Along with the shape and DNA sequence, other things such as their metabolic activities, conditions required for their growth, biochemical reactions, antigenic properties etc. are also helpful in classifying the bacteria.

Based on Morphology, DNA Sequencing, and Biochemistry

Based on the morphology, DNA sequencing, conditions required and biochemistry, scientists have come up with the following classification with 28 different bacterial phyla:

1. Acidobacteria
2. Actinobacteria
3. Aquificae
4. Bacteroidetes
5. Caldiserica
6. Chlamydiae
7. Chlorobi
8. Chloroflexi
9. Chrysiogenetes
10. Cyanobacteria
11. Deferribacteres
12. Deinococcus-Thermus
13. Dictyoglomi
14. Elusimicrobia
15. Fibrobacteres
16. Firmicutes
17. Fusobacteria
18. Gemmatimonadetes
19. Lentisphaerae

20. Nitrospira
21. Planctomycetes
22. Proteobacteria
23. Spirochaetes
24. Synergistetes
25. Tenericutes
26. Thermodesulfobacteria
27. Thermotogae
28. Verrucomicrobia

Each phylum further corresponds to the number of species and genera of bacteria. In a broad sense, this bacterial classification includes bacteria which are found in various types of environment such as fresh-water bacteria, saline-water bacteria, bacteria that can survive extreme temperatures (as in sulfur-water-spring bacteria and bacteria found in Antarctica ice), bacteria that can survive in highly acidic environment, bacteria that can survive in highly alkaline environment, bacteria that can withstand high radiations, aerobic bacteria, anaerobic bacteria, autotrophic bacteria, heterotrophic bacteria, and so on.

Though bacteria are mainly classified into phylum i.e. the scientific classification of organisms, they can be categorized into the following groups for simplification. Let's take a look.

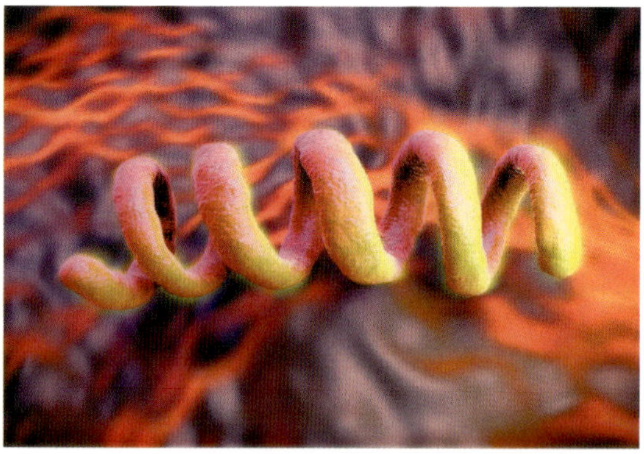

Spiral-shaped Bacteria (Treponema Pallidum)

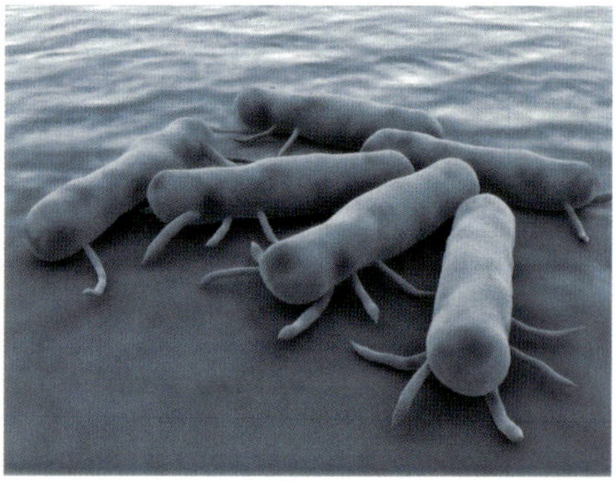

Rod-shaped Bacteria (Salmonella)

Before the advent of DNA sequencing, bacteria were classified based on their shapes and biochemical properties. Most of the bacteria can be classified into the following shapes.

Shapes of Bacteria

- Comma-shaped bacteria - e.g. Vibrio Cholera
- Spherical-shaped bacteria (Cocci) - e.g. Staphylococcus and Streptococcus
- Rod-shaped bacteria (Bacilli) - e.g. E.Coli and Salmonella
- Spiral-shaped bacteria (Spirilla) - e.g. Treponema and Borellia
- Flagellated bacteria - e.g. Tetanus Bacteria

Some bacteria have different, more complex shapes than the ones mentioned above.

Gram Positive and Gram Negative Bacteria

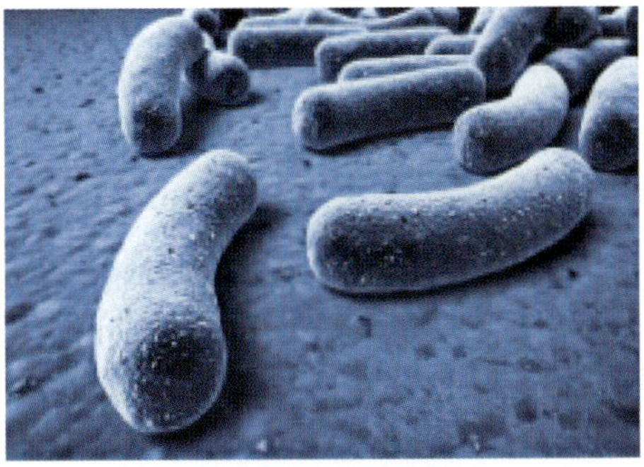

Gram Positive Bacteria

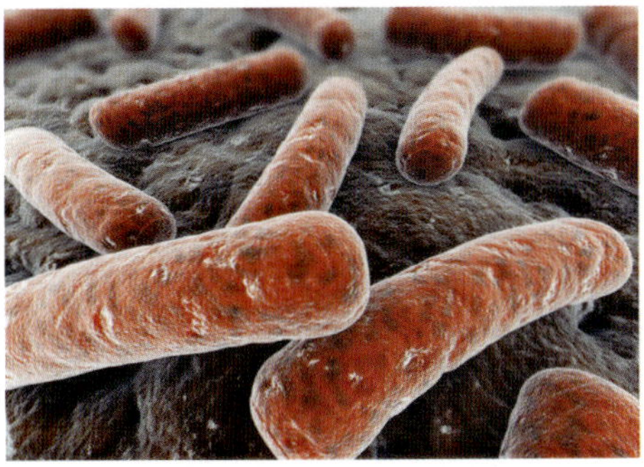

Gram Negative Bacteria

Bacteria are grouped as 'Gram positive' and 'Gram negative' bacteria, based on the results of Gram staining method. Peptidoglycans are the main contents of the cell walls of Gram-positive bacteria (almost 95%), while Gram-negative bacteria have an additional layer of phospholipids and lipopolysaccharides.

Gram-positive bacteria - The thick layer of Peptidoglycans is stained purple by the crystal violet dye, which is why gram-positive bacteria appear purple or blue.

Gram-negative bacteria - The thin layer of Peptidoglycans cannot retain the crystal violet dye, and thus appear red or pink due to the retention of the counter-stain.

Gram staining is a crucial method of bacterial identification, but there are a few bacteria which do not respond to it. Hence, two other groups are Gram-variable and Gram-indeterminate.

Based on Presence of Flagella

Some bacteria have whip-like structures on their bodies which help with motility, but that is not the only means to obtain motility. Flagellae are usually used for swimming.

- Flagellae absent - Atrichous Bacteria
- Flagellae present - These are furthur classified based on how many flagellae they possess and their location on the body.

1. Monotrichous Bacteria - One flagella
2. Amphitrichous Bacteria - One flagella on either side of the body
3. Polytrichous Bacteria - Multiple flagellae at different locations
4. Lopotrichous Bacteria - Flagellae only at one location on the body
5. Peritrichous Bacteria - Flagellae all over the body

Based on Requirement of Oxygen

Bacteria are also classified based on the requirement of oxygen for their survival.

- Aerobic bacteria - Bacteria that need oxygen for their survival.
- Anaerobic bacteria - Bacteria that do not require oxygen for survival.

Anaerobic bacteria cannot bear oxygen and may die if kept in an oxygenated environment. Such bacteria are usually found in places under the surface of the Earth or deep in oceans.

Based on Method of Obtaining Nutrition

This is one of the most important classification type, as it takes into account the most important aspect of bacterial growth and reproduction.

Photoautotrophic Bacteria

They synthesize their own food from organic matter, light energy and carbon-dioxide.

Chemoautotrophic Bacteria

They synthesize their food with the help of energy obtained from chemical sources.

Heterotrophic Bacteria

They obtain their food from other living organisms, as they cannot synthesize it on their own.

Symbiotic Bacteria

They obtain nutrition from host organism by offering something in return. Establish a mutual give-and-take relationship with host.

Pathogenic Bacteria

They obtain food from host but are harmful to the host, generally causing diseases.

Saprophytic Bacteria

They obtain nutrition from dead and decaying matter. Environmentally useful.

Based on Formation of Spores

Some bacteria form endospores, which are extremely tough and impenetrable outer shells, when exposed to unfavorable conditions. These endospores enable the bacteria to survive these conditions by remaining in a dormant state. When the conditions are favorable, the bacteria again revert to their original state. Endospores can help bacteria survive for millions of years in a dormant state.

Based on whether bacteria form endospores or not, they are classified into the following two types.

- Endospore forming bacteria
- Non-endospore forming bacteria

Bacteria which form endospores are generally more virulent in nature than those which don't.

Based on Environment

Various bacteria thrive in varied environment. While some species can withstand extreme conditions, others need specific moderate conditions to survive. Based on the preference of environmental conditions for their habitat, bacteria are classified into:

- Halophiles - Those which can survive in highly saline conditions.
- Thermophiles - Those which can resist high temperature.
- Acidophiles - Those which can tolerate low pH conditions.

- Neutrophiles - Those which require moderate conditions to survive.
- Mesophiles - Those which require moderate conditions to survive.
- Extremophiles - Those which can survive in extreme conditions.
- Alkaliphiles - Those which can tolerate high pH conditions.
- Psychrophilic bacteria - Those which can survive extremely cold conditions.
- Osmophiles - Those which can survive in high sugar osmotic conditions.

The great deal of diversity that you get to see in bacterial species makes it herculean task to come up with a single classification which will include every single species. Add to it the fact, that these bacteria are also constantly evolving and adapting to the environment wherein survival of living things has been deemed impossible (The best example are the bacteria which inhabit the brine lakes of Mediterranean Sea wherein the salinity levels far exceed the salinity levels of normal sea water.), and the task becomes a tad more difficult.

Diversity of Structure of Bacteria

Although bacterial cells are much smaller and simpler in structure than eukaryotic cells, the bacteria are an exceedingly diverse group of organisms that differ in size, shape, habitat, and metabolism. Much of the knowledge about bacteria has come from studies of disease-causing bacteria, which are more readily isolated in pure culture and more easily investigated than are many of the free-living species of bacteria. It must be noted that many free-living bacteria are quite different from the bacteria that are adapted to live as animal parasites or symbionts. Thus, there are no absolute rules about bacterial composition or structure, and there are many exceptions to any general statement.

Individual bacteria can assume one of three basic shapes: spherical (coccus), rodlike (bacillus), or curved (vibrio, spirillum, or spirochete). Considerable variation is seen in the actual shapes of bacteria, and cells can be stretched or compressed in one dimension. Bacteria that do not separate from one another after cell division form characteristic clusters that are helpful in their identification. For example, some cocci are found mainly in pairs, including Streptococcus pneumoniae, a pneumococcus that causes bacterial lobar pneumonia, and Neisseria gonorrhoeae, a gonococcus that causes the sexually transmitted disease gonorrhea. Most streptococci resemble a long strand of beads, whereas the staphylococci form random clumps (the name "staphylococci" is derived from the Greek word staphyle, meaning "cluster of grapes"). In addition, some coccal bacteria occur as square or cubical packets. The rod-shaped bacilli usually occur singly, but some strains form long chains, such as rods of the corynebacteria, normal inhabitants of the mouth that are frequently attached to one another at random angles. Some bacilli have pointed ends, whereas others have squared ends, and some rods are bent into a comma shape. These bent rods are often called vibrios and include Vibrio cholerae, which causes cholera. Other shapes of bacteria include the spirilla, which are bent and rebent, and the spirochetes, which form a helix similar to a corkscrew, in which the cell body is wrapped around a central fibre called the axial filament.

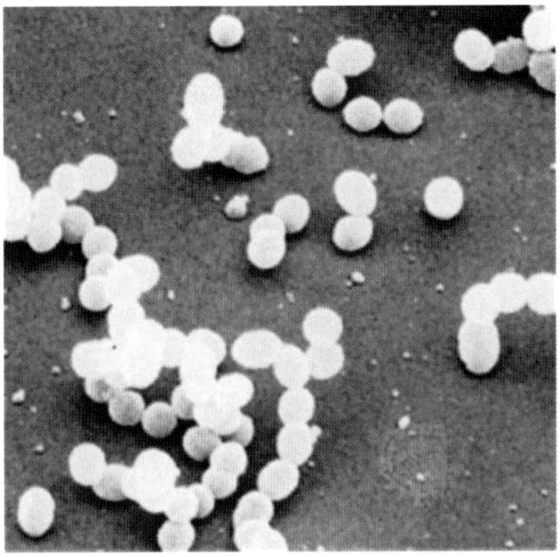

The bacterium Streptococcus mutans is an example of a spherical (coccus) bacterium. This species of bacteria commonly aggregates into pairs and short chains.
David M. Phillips/Visuals Unlimited

Bacteria are the smallest living entities. An average-size bacterium—such as the rod-shaped *Escherichia coli*, a normal inhabitant of the intestinal tract of humans and animals—is about 2 micrometres (μm; millionths of a metre) long and 0.5 μm in diameter, and the spherical cells of Staphylococcus aureus are up to 1 μm in diameter. A few bacterial types are even smaller, such as Mycoplasma pneumoniae, which is one of the smallest bacteria, ranging from about 0.1 to 0.25 μm in width and roughly 1 to 1.5 μm in length; the rod-shaped *Bordetella pertussis*, which is the causative agent of whooping cough, ranging from 0.2 to 0.5 μm in diameter and 0.5 to 1 μm in length; and the corkscrew-shaped *Treponema pallidum*, which is the causative agent of syphilis, averaging only 0.1 to 0.2 μm in diameter but 6 to 15 μm in length. The cyanobacterium Synechococcus averages about 0.5 to 1.6 μm in diameter. Some bacteria are relatively large, such as *Azotobacter*, which has diameters of 2 to 5 μm or more; and Achromatium, which has a minimum width of 5 μm and a maximum length of 100 μm, depending on the species. Giant bacteria can be visible with the unaided eye, such as Thiomargarita namibiensis, which averages 750 μm in diameter, and the rod-shaped Epulopiscium fishelsoni, which ranges from 30 to more than 600 μm in length.

Bacteria are unicellular microorganisms and thus are generally not organized into tissues. Each bacterium grows and divides independently of any other bacterium, although aggregates of bacteria, sometimes containing members of different species, are frequently found. Many bacteria can form aggregated structures called biofilms. Organisms in biofilms often display substantially different properties from the same organism in the individual state or the planktonic state. Bacteria that have aggregated into biofilms can communicate information about population size and metabolic state. This type of communication is called quorum sensing and operates by the production of small molecules called autoinducers or pheromones. The concentration of quorum-sensing molecules—most commonly peptides or acylated homoserine lactones (AHLs; special signaling chemicals)—is related to the number of bacteria of the same or different species that are in the biofilm and helps coordinate the behaviour of the biofilm.

Bacteria cells function quite differently from human cells, from the function of the cell as a whole, right down to the individual structures in each cell. Bacteria exist in nature as individual cells. While some bacteria work together in a group, no bacteria forms multicellular tissues. Bacteria have organized structures within the cell, but do not have membrane-bound organelles, as with human cells.

Bacterial Cell Function

In the grand scheme of things, the function of each bacterial cell begins and ends with collecting enough nutrients to survive. Bacterial cells consist of a phospholipid bilayer, and in some cases a layer of peptidoglycan. The phospholipid bilayer either allows or denies ions and other molecules entry to the cell, based on factors such as size and charge. The peptidoglycan, while not technically a cell wall, serves a similar function by protecting against the osmotic gradient created across the phospholipid bilayer.

Genetic Material

In human cells, deoxyribonucleic acid (DNA) gets produced in the membrane-bound nucleus. DNA from bacterial cells resides throughout the interior of the cell, in the cytoplasm. The main function of DNA in bacterial cells is the same as human cells, transcription into ribonucleic acid (RNA) followed by translation into amino acids and subsequent folding into proteins.

Ribosomes

Ribosomes exist in the cytoplasm of a bacterial cell and have a very important function. Ribosomes translate the amino acids from RNA. Once translated, the amino acids fold up into tertiary structures called proteins. The proteins in a bacterial cell have many important functions, including regulation of transcription and translation, and becoming transmembrane proteins for help in relocation of necessary nutrients that are too large or charged to pass across the phospholipid bilayer.

Flagella

Flagella help bacteria move through the environment. The flagellum is a whip-like structure that can occur at one end, both ends, or all over the bacterial cell. The whipping action pushes the bacteria through the aqueous solution and improves the cells' chances of being in a nutrient-rich environment.

Endospores

Some species of bacteria contain an endospore. Endospores are a bacteria cell's way of protecting itself against harsh changes in the environment or nutrient depletion. When a required nutrient in the environment becomes depleted or absent, signals get sent to the bacteria to begin endospore formation. An endospore protects the bacterial genetic material so that, when optimal conditions return, the bacterial cell can reform and thrive again.

The Importance of Bacteria to Humans

Bacteria in Food

Milk from a healthy cow initially contains very few bacteria, which primarily come from the skin of the cow and the procedures for handling the milk. Milk is an excellent growth medium for numerous bacteria, and the bacteria can increase rapidly in numbers unless the milk is properly processed. Bacterial growth can spoil the milk or even pose a serious health hazard if pathogenic bacteria are present. Diseases that can be transmitted from an infected cow include tuberculosis (Mycobacterium tuberculosis), undulant fever (Brucella abortus), and Q fever (Coxiella burnetii). In addition, typhoid fever (Salmonella typhi) can be transmitted through milk from an infected milk handler. Pasteurization procedures increase the temperature of the milk to 63 °C (145 °F) for 30 minutes or to 71 °C (160 °F) for 15 seconds, which kills any of the pathogenic bacteria that might be present, although these procedures do not kill all microorganisms.

Certain bacteria convert milk into useful dairy products, such as buttermilk, yogurt, and cheese. Commercially cultured buttermilk is prepared from milk inoculated with a starter culture of Lactococcus (usually L. lactis or L. lactis cremoris). Yogurt and other fermented milk products are produced in a similar manner using different cultures of bacteria. Many cheeses are likewise made through the action of bacteria. Growth in milk of an acid-producing bacterium such as L. lactis causes the casein to precipitate as curd. Following the removal of moisture and the addition of salt, the curd is allowed to ripen through the action of other microorganisms. Different bacteria impart different flavours and characteristics to foods; for example, the mixture of *Lactobacillus casei*, Streptococcus thermophilus, and Propionibacterium shermanii is responsible for the ripening of Swiss cheese and the production of its characteristic taste and large gas bubbles. In addition, Brevibacterium linens is responsible for the flavour of Limburger cheese, and molds (Penicillium species) are used in the manufacture of Roquefort and Camembert cheeses. Other types of bacteria have long been used in the preparation and preservation of various foods produced through bacterial fermentation, including pickled products, sauerkraut, and olives.

The toxins of many pathogenic bacteria that are transmitted in foods can cause food poisoning when ingested. These include a toxin produced by *Staphylococcus aureus*, which causes a rapid, severe, but limited gastrointestinal distress, or the toxin of *Clostridium botulinum*, which is often lethal. Production of botulism toxin can occur in canned nonacidic foods that have been incomplctcly cooked before sealing. C. botulinum forms heat-resistant spores that can germinate into vegetative bacterial cells that thrive in the anaerobic environment, which is conducive to the production of their extremely potent toxin. Other food-borne infections are actually transmitted from an infected food handler, including typhoid fever, salmonellosis (Salmonella species), and shigellosis (Shigella dysenteriae).

Substances contributing to the virulence of pathogenic bacteria	
substance	**action**
hyaluronidase	increases permeability of tissue spaces to bacterial cells
coagulase	increases resistance of bacteria to phagocytosis (engulfment by defense cells, or phagocytes)
hemolysins	destroy red blood cells
collagenase	dissolves collagen, a connective tissue protein
leukocidin	kills white blood cells (specifically leukocytes) and hence decreases phagocytic action
exotoxins and endotoxins	interfere with normal metabolic processes

Virus

Virus, an infectious agent of small size and simple composition that can multiply only in living cells of animals, plants, or bacteria. The name is from a Latin word meaning "slimy liquid" or "poison."

The earliest indications of the biological nature of viruses came from studies in 1892 by the Russian scientist Dmitry I. Ivanovsky and in 1898 by the Dutch scientist Martinus W. Beijerinck. Beijerinck first surmised that the virus under study was a new kind of infectious agent, which he designated contagium vivum fluidum, meaning that it was a live, reproducing organism that differed from other organisms. Both of these investigators found that a disease of tobacco plants could be transmitted by an agent, later called tobacco mosaic virus, passing through a minute filter that would not allow the passage of bacteria. This virus and those subsequently isolated would not grow on an artificial medium and were not visible under the light microscope. In independent studies in 1915 by the British investigator Frederick W. Twort and in 1917 by the French Canadian scientist Félix H. d'Hérelle, lesions in cultures of bacteria were discovered and attributed to an agent called bacteriophage ("eater of bacteria"), now known to be viruses that specifically infect bacteria.

The unique nature of these organisms meant that new methods and alternative models had to be developed to study and classify them. The study of viruses confined exclusively or largely to humans, however, posed the formidable problem of finding a susceptible animal host. In 1933 the British investigators Wilson Smith, Christopher H. Andrewes, and Patrick P. Laidlaw were able to transmit influenza to ferrets, and the influenza virus was subsequently adapted to mice. In 1941 the American scientist George K. Hirst found that influenza virus grown in tissues of the chicken embryo could be detected by its capacity to agglutinate (draw together) red blood cells.

A significant advance was made by the American scientists John Enders, Thomas Weller, and Frederick Robbins, who in 1949 developed the technique of culturing cells on glass surfaces; cells could then be infected with the viruses that cause polio (poliovirus) and other diseases. (Until this time, the poliovirus could be grown only in the brains of chimpanzees or the spinal cords of monkeys.) Culturing cells on glass surfaces opened the way for diseases caused by viruses to be identified by their effects on cells (cytopathogenic effect) and by the presence of antibodies to them in the blood. Cell culture then led to the development and production of vaccines (preparations used to elicit immunity against a disease) such as the poliovirus vaccine.

Scientists were soon able to detect the number of bacterial viruses in a culture vessel by measuring their ability to break apart (lyse) adjoining bacteria in an area of bacteria (lawn) overlaid with an inert gelatinous substance called agar—viral action that resulted in a clearing, or "plaque." The American scientist Renato Dulbecco in 1952 applied this technique to measuring the number of animal viruses that could produce plaques in layers of adjoining animal cells overlaid with agar. In the 1940s the development of the electron microscope permitted individual virus particles to be seen for the first time, leading to the classification of viruses and giving insight into their structure.

Advancements that have been made in chemistry, physics, and molecular biology since the 1960s have revolutionized the study of viruses. For example, electrophoresis on gel substrates gave a deeper understanding of the protein and nucleic acid composition of viruses. More-sophisticated immunologic procedures, including the use of monoclonal antibodies directed to specific antigenic sites on proteins, gave a better insight into the structure and function of viral proteins. The progress made in the physics of crystals that could be studied by X-ray diffraction provided the high resolution required to discover the basic structure of minute viruses. Applications of new knowledge about cell biology and biochemistry helped to determine how viruses use their host cells for synthesizing viral nucleic acids and proteins.

The revolution that took place in the field of molecular biology allowed the genetic information encoded in nucleic acids of viruses—which enables viruses to reproduce, synthesize unique proteins, and alter cellular functions—to be studied. In fact, the chemical and physical simplicity of viruses has made them an incisive experimental tool for probing the molecular events involved in certain life processes. Their potential ecological significance was realized in the early 21st century, following the discovery of giant viruses in aquatic environments in different parts of the world.

General Features

Viruses occupy a special taxonomic position: they are not plants, animals, or prokaryotic bacteria (single-cell organisms without defined nuclei), and they are generally placed in their own kingdom. In fact, viruses should not even be considered organisms, in the strictest sense, because they are not free-living; i.e., they cannot reproduce and carry on metabolic processes without a host cell.

All true viruses contain nucleic acid—either DNA (deoxyribonucleic acid) or RNA (ribonucleic acid)—and protein. The nucleic acid encodes the genetic information unique for each virus. The infective, extracellular (outside the cell) form of a virus is called the virion. It contains at least one unique protein synthesized by specific genes in the nucleic acid of that virus. In virtually all viruses, at least one of these proteins forms a shell (called a capsid) around the nucleic acid. Certain viruses also have other proteins internal to the capsid; some of these proteins act as enzymes, often during the synthesis of viral nucleic acids. Viroids (meaning "viruslike") are disease-causing organisms that contain only nucleic acid and have no structural proteins. Other viruslike particles called prions are composed primarily of a protein tightly complexed with a small nucleic acid molecule. Prions are very resistant to inactivation and appear to cause degenerative brain disease in mammals, including humans.

Viruses are quintessential parasites; they depend on the host cell for almost all of their life-sustaining functions. Unlike true organisms, viruses cannot synthesize proteins, because they lack ribosomes (cell organelles) for the translation of viral messenger RNA (mRNA; a complementary copy of the nucleic acid of the nucleus that associates with ribosomes and directs protein synthesis) into proteins. Viruses must use the ribosomes of their host cells to translate viral mRNA into viral proteins.

Viruses are also energy parasites; unlike cells, they cannot generate or store energy in the form of adenosine triphosphate (ATP). The virus derives energy, as well as all other metabolic functions, from the host cell. The invading virus uses the nucleotides and amino acids of the host cell to synthesize its nucleic acids and proteins, respectively. Some viruses use the lipids and sugar chains of the host cell to form their membranes and glycoproteins (proteins linked to short polymers consisting of several sugars).

The true infectious part of any virus is its nucleic acid, either DNA or RNA but never both. In many viruses, but not all, the nucleic acid alone, stripped of its capsid, can infect (transfect) cells, although considerably less efficiently than can the intact virions.

The virion capsid has three functions: (1) to protect the viral nucleic acid from digestion by certain enzymes (nucleases), (2) to furnish sites on its surface that recognize and attach (adsorb) the virion to receptors on the surface of the host cell, and, in some viruses, (3) to provide proteins that form part of a specialized component that enables the virion to penetrate through the cell surface membrane or, in special cases, to inject the infectious nucleic acid into the interior of the host cell.

Host Range and Distribution

Logic originally dictated that viruses be identified on the basis of the host they infect. This is justified in many cases but not in others, and the host range and distribution of viruses are only one criterion for their classification. It is still traditional to divide viruses into three categories: those that infect animals, plants, or bacteria.

Virtually all plant viruses are transmitted by insects or other organisms (vectors) that feed on plants. The hosts of animal viruses vary from protozoans (single-celled animal organisms) to humans. Many viruses infect either invertebrate animals or vertebrates, and some infect both. Certain viruses that cause serious diseases of animals and humans are carried by arthropods. These vector-borne viruses multiply in both the invertebrate vector and the vertebrate host.

Certain viruses are limited in their host range to the various orders of vertebrates. Some viruses appear to be adapted for growth only in ectothermic vertebrates (animals commonly referred to as cold-blooded, such as fishes and reptiles), possibly because they can reproduce only at low temperatures. Other viruses are limited in their host range to endothermic vertebrates (animals commonly referred to as warm-blooded, such as mammals).

Types of virus

These pathogens are one of the most widespread of all organisms and are capable of infecting every species of animal from mammals down to insects, protozoa, and even bacteria and plants. In fact there are more species of virus than of all other creatures put together.

Some are harmless while some are extremely dangerous and like HIV causing AIDS or Ebola and Marburg virus etc.

Origin of viruses

Researchers have found resemblance of the genomes of viruses with genes of higher animals that can 'jump' from one chromosome to another. Some believes these viruses may have originated from bacterial plasmids, which are little packets of genes lying outside the bacterial chromosomes and capable of being transferred to another bacterium.

Viruses may also have originated from degenerate bacteria that have become obligate parasites.

Viral Structure, Replication, and Function

Viruses vary greatly in size as well as complexity. But some of their features are common among all species of viruses. There are three basic parts:

- nucleic acid
- protein coat
- lipid membrane

Viral Nucleic Acid

The nucleic acid is the core of the virus. It is either DNA or RNA (deoxyribonucleic acid and ribonucleic acid respectively). The DNA or RNA holds all of the information for the virus and that makes it unique and helps it multiply.

There may be either RNA or DNA and never both and this determines the way in which viruses replicate themselves. Both types of genome however function to make viral proteins and more copies of the viral coat and DNA/RNA.

DNA viruses are the simplest. They use the host cell's RNA polymerase can make mRNA that translate on the host ribosomes to make viral proteins.

RNA viruses on the other hand need to provide their own RNA polymerase to make mRNA.

In addition, those that are negative-sense RNA viruses need yet another step to make positive-sense RNA.

Retroviruses need another enzyme that they carry with them called reverse transcriptase. This converts RNA to DNA that is inserted into the host genome. Then the synthesis of viral proteins, and their assembly into new viral particles may take place in the host cell's nucleus (in influenza virus, measles virus) or in the cytoplasm (e.g. rabies, herpes).

Protein Coat

This is a covering over the nucleic acid that protects it. This has a symmetrical structure and is built of one or more subunits packed like a chemical crystal.

Lipid Membrane

This covers the capsid. Many viruses do not have this envelope and are called naked viruses. This membrane is usually acquired by the virus from the host cell in the process of leaving the cell. This coat enables the virus to survive outside the cell sufficiently long to spread elsewhere via the blood.

Viral Structure

The capsid and entire virus structure can be of four main types:

1. Helical

There is a capsomer coiled around a central axis to form a helical structure. This is a common structure seen in single stranded RNA viruses. Tobacco mosaic virus is a helical virus.

2. Icosahedral

These are near-spherical and this shape is adopted because the coat forms a closed shell. Rota virus has twelve capsomers and appear spherical.

3. Envelope

The virus is covered with a lipid membrane in a modified form of one of the cell membranes. The outer membrane is from the infected host cell and internal membranes from nuclear membrane or endoplasmic reticulum forming a lipid bilayer known as a viral envelope. This membrane is studded with proteins or receptors.

4. Complex

There is a capsid that is neither purely helical, nor purely icosahedral. There may be extra features like protein tails or a complex outer wall. Bacteriophages are examples of this type of viral structure.

Viral Receptors

The virus has to enter the cell in order to infect it. Viruses are not taken up by the cell directly. They must attach to a receptor on the cell surface first in order to gain entry. If the receptor is not a necessary one, the cell once infected with the virus, may go on to remove the receptor altogether.

Each virus has its specific receptor and it is a vital component of the cell surface so that the cell cannot get rid of it to avoid the infection. The selectivity of the viruses determines the cell-preference.

For example, rhinoviruses have a preference for cells lining the nose, airways and the lungs, HIV with CD4, CCR5, CXCR4 viruses, Epstein–Barr virus (EBV) with Rabies virus with CR2, Acetylcholine receptor, Influenza virus with Neuraminic acid on red blood cells etc.

HIV infects mainly T lymphocytes and macrophages because only they carry a surface molecule known as CD4 receptor and EBV infects B lymphocytes carrying the complement receptor CR2.

Protozoa

Organisms known as protozoa include a wide range of organisms, most of which are free-living single-celled eukaryotes. Therefore, protozoa fit into the Domain Eukarya. Although the different phyla of the kingdom Protista are not closely related, they are nonetheless classified together because of their large differences from the other kingdoms of plants, animals and fungi. The name "protozoa" has a dynamic history, at one time including only the "animal-like" unicellular forms of life. Today, these heterotrophic protozoa are lumped together with the autotrophic algae and other simple forms of life into the Kingdom Protista. While all protozoa are eukaryotes, not all reproduce with the standard model of mitosis that is seen in higher animal cells. Many have complex cellular division that resembles binary fission in bacteria, on a larger scale. Some phyla in the Kingdom Protista are autotrophic cells, containing chloroplasts which can produce sugars from sunlight. Although only heterotrophic organisms were considered protozoa historically. Below is an image of a ciliate protozoa.

Examples of Protozoa

Malaria

Malaria is a disease that effects hundreds of millions of people worldwide, every year. It is estimated that malaria kills three million people a year, half of which are young children. Malaria is caused by several protozoa in the genus Plasmodium. This particular protozoa carries out life cycles in both the digestive systems of mosquitos, and in human blood cells. The unicellular protozoans enter the bloodstream of humans during a mosquito bite. They find an uninfected red blood cell, and use a special combination of organelles at the end of their cell to force the cell to draw the parasite in. This special group of organelles, known as the apical complex, allows the parasitic protozoa to hide within blood cells. This makes Plasmodium and other parasitic Apicomplexans very hard to kill, as the blood cells would have to be destroyed as well.

Red Tide

An interesting phenomena in the ocean is also caused by another phylum of protozoa. The Dinoflagellates are a group of protozoa that exist mainly as free-swimming unicellular organisms, while other exist symbiotically within certain invertebrates. Groups of dinoflagellates that are free swimming can undergo drastic periods of reproduction, known as blooms. If a bloom produces enough dinoflagellates, the ocean water will become pink or even red because of their density. Unfortunately for any organism that passes through the bloom, dinoflagellates produce a toxin. When congregated in enormous densities, this neurotoxin can paralyze and kill almost any organism in the bloom. Fish, crustaceans, and mollusks that eat the bloom may store the toxins in their tissues and become poisonous to people who eat them. Paralytic Shellfish Poisoning is the caused by dinoflagellate toxins in high concentration. Simple protozoa, in high numbers, can cause muscle paralysis and respiratory failure in humans, possibly leading to death.

Characteristics of Protozoa

Protozoa do not have a cell wall and therefore can have a variety of shapes. Nevertheless, some of the protozoans have a pliant layer, a pellicle, or a stiff shell outside the cell membrane.

Protozoa vary in size and shape. Their sizes range from 10 to 55 micrometers, but they can be as large as 1 mm. The largest protozoa are called xenophyophores, which can measure up to 20 centimeters in diameter.

Protozoa prefer living in moist and aquatic habitats. Their cysts can be found in the bleakest parts of the ecosphere.

Protozoa are found drifting in the oceans, seas, and freshwater. They are at the base of food chains.

The life cycle of protozoa changes between proliferative stages and dormant cysts.

When in the cystic stage, protozoa can live in utmost temperatures or harsh chemicals, or without nutrients, water, or oxygen for a long time. Being a cyst enables parasitic species to dwell on the host externally. This lets them transmit from one host to another. In the form of trophozoites, protozoa feed actively. The transition of a trophozoite to a cyst is called encystation and the transition back to a trophozoite is called excystation.

The mode of nutrition of protozoa is heterotrophic, and most species obtain food by phagocytosis. Phagocytosis is the process where the cell changes shape by sending out pseudopodia to make contact with food particles.

Protozoa take food into the cell at a point called the cytostome. The food is ingested by them and lysosomal enzymes digest the food. There are also certain types of protozoa that take in food by their cell membranes. Some others such as the amoeba, surround food and absorb it. Others have mouth pores into which they pull in food.

Protozoans digest their food in spaces called vacuoles. Contractile vacuoles that are found in protozoa thriving in freshwater, excrete water that penetrates into the cells by osmosis. While chewing down the food, protozoans produce and release nitrogen.

Protozoa species move on their own by one of the three types of locomotor organelles such as flagella, cilia, or pseudopodia.

Protozoa reproduce by the method of binary fission or multiple fission. Some of the members reproduce by asexual mode, some by sexual means, and some by both.

Classification

The protozoa group comprises more than 65,000 species. All the protozoan species belong to the kingdom Protista. Many kinds of protozoa are symbionts. Some of the protozoan species are parasites and some are predators of bacteria and algae. Some examples of protozoans are dinoflagellates, amoebas, paramecia, and plasmodium.

Types of Protozoa

There are many phyla in the Domain Protista. While this represents a wide variety of them, it is not close to all of them. Part of the reason there is so much diversity in this domain is that the phyla are not closely related.

Phylum Euglenida

The Euglenida are a phylum of protozoa recognized by the pellicle that gives them shape and the flagella which they use for locomotion. The pellicle is a shell of sorts that exists beneath the cell membrane. It is composed of strips of proteins, which interlock for support. Some organisms in the Euglenida are photosynthetic, and contain chloroplasts. Others obtain food from dissolved nutrients in the environment, while still others are parasitic.

Phylum Kinetoplastida

Closely related to the Euglenida, the Kinetoplastida are also protected by a pellicle, although it is made exclusively of microtubules. Organisms in the Kinetoplastida share the unique characteristic of having a single, much enlarged and elongated, mitochondrion. Typically, cells have many small mitochondria, as opposed to one large one. The Kinetoplastida includes many parasitic organisms that cause disease in humans. Of these, leishmaniosis is the most notable, affecting over a million people a year. However, advances in medical treatments saves most of those infected, and only around 1000 people die annually from these protozoa.

Phylum Ciliophora

Instead of flagella for locomotion, organisms in the phylum Ciliophora use much smaller structures called cilia. The cilia of these organisms cover their entire cell, and work together to propel the cell forward. Much like the individual paddles of a row boat, each cilia gives a forward moving power stoke, then whips back to the starting position in the recovery stroke. Organisms in the Ciliophora include a wide variety of body plans, including free-swimming organisms and sessile organisms that use their cilia to filter food from the water. Most ciliates exist on the bottom of marine environments, known as the benthic zone. However, these protozoa have also specialized as parasites in the digestive tracts of larger organisms.

Phylum Apicomplexa

The Apicomplexa include mostly parasitic organisms that exist solely within their host organisms. These organisms all have apical complexes, which allow them to gain entry into their host's cells. Inside the cells, the parasites are much more resistant and have better access to nutrients. Many of these parasites can also change the proteins exposed on the surface of their cells. This helps them "hide" from the immune system cells, which recognize foreign cells on the surface of invading pathogens. This makes it very difficult to treat medically.

Phylum Dinoflagellata

The dinoflagellates use flagella to locomote and are also covered by a pellicle. Unlike the two previous groups with pellicles, the dinoflagellates use a series of vesicles beneath the cell membrane to create a rigid form. Some dinoflagellates even fill their vesicles with polysaccharides, like cellulose, to create a form of armor that protects their cells.

Phylum Stramenopila

The Stramenopila includes a wide variety of organisms, from the shelled diatoms to brown and golden algae. Protozoa in this phylum have shells, scales, or tests that support the cell. These protective covering vary widely across the phylum. The diatoms tend to make ornate and complex tests of silicate, which is also used to make glass. Others form their shells from calcium carbonate or protein.

Phylum Rhizopoda

The phylum Rhizopoda contains the amoebas. The small, unicellular protozoa are some of the only protozoa that do not have any sort of hard covering. The amoebas move by extending their cytoplasm into the environment. These extensions are called pseudopodia.

Phylum Actinopoda

The organisms in Actinopoda have characteristic axopodia, which are sharp spines that extend from the cell and are covered in cytoplasm. They are used in locomotion and feeding. The axopodia also produce beautiful shapes, as seen in the organism below:

Radiolarian

Phylum Granuloreticulosa

The Granuloreticulosa includes a variety of organism that have immense industrial value. The tests produced by these protozoa aggregate in huge numbers on the bottom of marine environments. Over time they fossilize together and become substances like chalk, limestone, and marble. Even the pyramids of Egypt were built from stones that originated from the shells of these protozoans.

Phylum Diplomonodida

The Diplomonodida are a phylum of organisms that have many flagella for locomotion. The average number is around 8. While there are many species that vary in their place in the ecosystem, the most well-known of these protozoa is the genus Giardia, which can be consumed in unclean water and cause flu-like symptoms and diarrhea in humans.

Phylum Parabasilida

These organisms contain many flagella, up to thousands on a single cell, and have a special fiber that attached the Golgi apparatus to the base of the flagella. Many parabasilids exists as symbiotic protozoa in the digestive tracts of insects, particularly those that eat wood. These protozoa are able to secrete an enzyme that can break down cellulose.

Other Protozoa

Several other phyla of protozoa exist, but cannot be covered here. Like all other protozoans, these organisms lack specialized tissues or layers of cells. Most are unicellular, although some exist in colonies or fibers. Due to their small size an adaptability, the protozoa have occupied nearly every environment on the planet.

Protozoan Disease-Amoebiasis

Amoebiasis, also known as amoebic dysentery, is caused by Entamoeba histolytica. Infection generally occurs through drinking water. The trophozoite of E. histolytica penetrates the wall of the colon, secretes histolytic enzymes and feeds upon its cells causing ulcers.

These ulcers rupture and discharge mucus and blood into the intestine that pass along with stools and results in amoebic dysentery. If the infection is allowed to continue the parasite may reach the liver, lungs and brain where it causes abscesses which prove fatal.

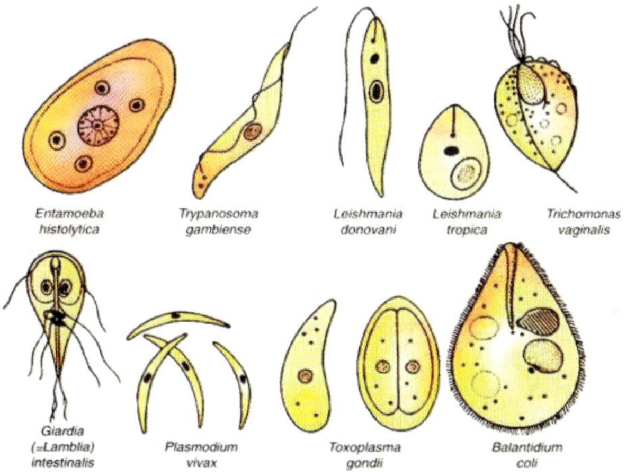

some pathogenic protozoan of man

There is no intermediate host in the life cycle of E. histolytica. Transmission of the parasite from man to man takes place through the tetra nucleate cysts. Before the cyst-formation the trophozoite changes into a smaller minuta form, which then encysts to form a tetra nucleate cyst.

These tetra nucleate cysts are voided with the faecal and contaminated water and food and are then transmitted into new hosts. Faecal contamination of drinking water, vegetables and food are the primary causes. Eating of uncooked vegetables and fruits which have been fertilised with infected human faeces has often led to the occurrence of disease.

Occasionally drinking water supply contaminated with infected faeces gives rise to epidemics. Houseflies may transmit cysts while passing from faeces to unprotected foodstuffs. The cysts of E. histolytica have been found in the droppings of cockroaches which also serve as a source of infection.

Amoebiasis is endemic in tropical countries. It can be treated with Emetine, Fumagillin, Metronidazole, Tinidazole, Terramycin, Erythromycin, Aureomycin and Chloroquine, etc.

Protozoan Disease-Diarrhoea

Diarrhoea, which is characterised by loose bowels, is caused by a flagellete parasite Giardia (=Lamblia) intestinalis and a sporozoan Isopora hominis. Giardia intestinalis is a parasite in the small intestine and colon of man.

It has an elliptical or pear-shaped body which is bilateraly symmetrical with dorsal side convex and ventral side flattened and deepened anteriorly to form a concave sucking disc. It bears two nuclei and four pairs of long flagella arranged symmetrically.

With the help of sucking disc the parasite attaches itself on to the convex surface of the epithelial cells in the intestine and may cause a disturbance of intestinal function leading to malabsorption of fat which causes diarrhoea. Consequently the patient may complain of persistent looseness of bowels. The parasite is also capable of producing epigastric pain, abdominal discomfort, loss of appetite, headache and toxic effect (allergy).

Infection in man is brought about by ingestion of cysts. Transmission of the parasite takes place through cysts which are voided with faeces and are transmitted to new hosts with contaminated water and food. The infection of Giardia is more common in children than in adults.

Atebrin, Chloroquine and Acranil are effective drugs in the treatment of Giardia. Metronidazole has also been reported to be quite effective in its treatment.

Isopora hominis is a rare infectious parasite amongst human beings inhabiting the small intestine of man and may cause intestinal symptoms like colic and diarrhoea. Dogs are supposed to be reservoir hosts of I. hominis.

Chapter 2

Food Spoilage Caused by Microbial Growth

The edibility of food is reduced considerably with time due to spoilage caused by microbial growth. The chapter closely examines the causative factors of food spoilage especially due to microbial growth. It includes essential topics relevant to the understanding of this subject, such as growth factors of microorganisms, meat spoilage and wine fault.

Growth Factors of Microorganisms

A characteristic of microorganisms is their ability to grow and form a population of organisms. One of the results of microbial metabolism is an increase in the size of the cell. The many requirements for successful growth include those both chemical and physical.

Following are the factors:

Solutes and Water Acidity

Water is one of the most essential requirements for life. Thus, its availability becomes most important factor for the growth of microorganisms. The availability of water depends on two factors — the water content of the surrounding environment and the concentration of solutes (salts, sugars, etc.) dissolved in the water.

In most cases, the cell cytoplasm possesses higher solute concentration in comparison to its environment. Thus, water always diffuses from a region of its higher concentration to a region of the lower concentration.

This process is called osmosis which keeps the microbial cytoplasm in positive water balance. When a microbial cell is placed in hypertonic solution (or, solution of low water activity) it loses water and shrinkage of membrane takes place. This phenomenon is called plasmolysis.

Microorganisms show variability in their ability to adapt the habitats of low water activity. Microorganisms like S. aureus can survive over a wide range of water activity and are called as osmotolerant (as water activity is inversely related to osmotic pressure).

However, most microorganisms grow well only near pure water activity (i.e., around 0.98-1). Thus, drying of food or addition of high concentration of salts and sugars is the most popular way of preventing spoilage of food.

Seawater microorganisms are called as halophiles since they require high concentration of salts (between 2.8-6.2 M) to grow. Halobacterium, a halophilic archaebacterium, inhabits Dead Sea (a salt lake situated between Israel and Jordan and the lowest lake in the world), the Great Salt Lake in Utah and other aquatic habitats possessing salt concentrations approaching salt water.

The quantitative availability of water can be expressed in physical term called water activity (a_w). The water activity for a sample solution is the ratio of vapour pressure of the sample solution to the vapour pressure of the water at the same temperature ($a_w = P_{soln}/P_{water}$).

The relative humidity of a test sample (at equilibrium) can be obtained, after sealing it in a closed chamber. This determines the water activity of a solution. For example, if after treating by above method relative humidity of air over the sample is 95% then, the water activity of the sample is 0.95.

Temperature

All forms of life are greatly influenced by temperature. In fact, the microorganisms are very sensitive to the temperature since their temperature varies with that of environment (puikilothermic).

Temperature influences the rate of chemical reactions and protein structure integrity thus affecting rates of enzymatic activity. At low temperature enzymes are not denatured, therefore, every 10°C rise in temperature results in rise of metabolic activity and growth of microorganisms.

However, enzymes have a range of thermal stability and beyond it their denaturation takes place. Thus, high temperature kills micro-organism by denaturing enzymes, by inhibiting transport carrier molecules or by change in membrane integrity. Each microbe shows characteristic temperature dependence and possesses its own cardinal temperatures i.e. minimal, maximum and optimal growth temperatures.

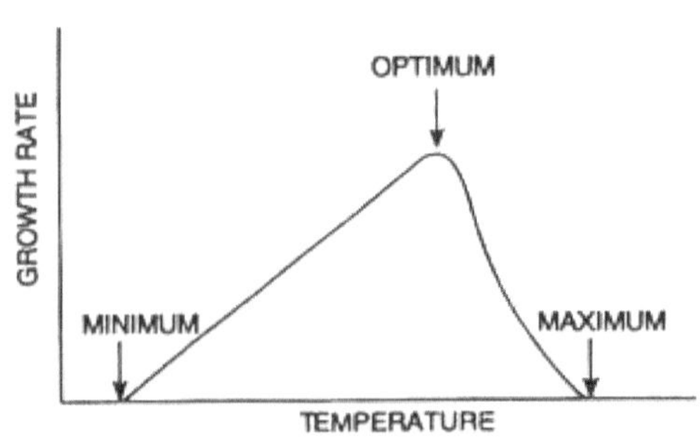

Cardinal temperature

The values for cardinal temperature vary widely among bacteria. For convenience, bacteria isolated from hot springs can survive even at temperature of 100°C and above, while those isolated from snow can survive below -10°C. On the basis of susceptibility to the thermal conditions, microorganisms are classified into three categories: thermophiles, meshophiles and psychrophiles.

Thermophiles are microorganisms that show growth optima at 55°C. They often have-growth maxima of 65°C, while few can grow even at 100°C and higher temperatures. Their growth minima are 45°C. The vast majority of thermophiles belong to prokaryotes although a few microalgae (e.g., Cyanidium caldarium) and micro fungi (e.g. Mucor pusillus) are also thermophiles.

A few microorganisms are hyperthermophiles as they possess growth optima between 80°C and about 113°C. Hyper-thermophiles usually do not grow well below 55°C (e.g., Pyrococcus abyssi, Pyrodictium occultum).

Mesophiles are microorganisms that have growth minima between 15°C-20°C; optima between 20-45°C and maxima at 45°C. Most microorganisms fall within this category. Almost all human pathogens are mesophiles as they grow at a fairly constant temp, of 37°C. Psychrophiles have optimum temperature for growth at 15°C, however, few can grow even below 0°C.

The maximum growth temperature of psychrophiles is around 20°C. The spoilage of refrigerated food takes place because of facultative psychrophiles. These are microorganisms that can grow at 0°C but have growth optima temperature between 20°-30°C.

pH

The pH is defined as a negative logarithm of hydrogenion concentration:

$$pH = \log \frac{1}{\left[H^+\right]}$$

The pH scale extends from pH 0.0 to pH 14.0 and each pH unit represents a tenfold change in hydrogen ion concentration. The pH of a solution reveals whether the solution is acidic, alkaline or neutral. A neutral solution contains pH equal to 7. The acid solution shows pH below 7, while alkaline solution exhibit pH above 7.

The activity of microbial enzymes depends on the change present on the surface of amino acids. Any change in the environmental pH may either enhance the enzyme activity or inhibit the activity.

Thus, pH can dramatically affect the growth of microorganisms. Each species of microorganisms shows specific pH growth range. Microorganisms can be classified as acidophilus, neutrophils and basophiles (alkalophiles) on the basis of their requirement for particular pH in their environment.

pH range of few bacteria			
Organisms	Minimum	Optimum	Maximum
Thiobacillus thiooxidans	1.0	2.0-2.8	4.0-6.0
Lactobacillus acidophilus	4.5-4.6	5.8-6.6	6.8
Clostridium sporogens	5.0-5.8	6.0-7.6	8.5-9.0
Pseudomonas aeroginosa	5.6	6.6-7.0	8.0
Nitrobacter sp.	6.6	7.6-8.6	10.0
Nitrosomonas sp.	7.0-7.6	8.0-8.8	9.4

The acidophiles grow between the pH range of 0.0 to 5.5, neutrophils between 5.5 to 8 and basophiles between 4.5 to 11.5. Most micro fungi are acidophiles as they grow in surroundings having pH about 4 to 6. Most bacteria and protozoa are neutrophils.

Growing microorganisms produce acidic and basic metabolic waste products. These wastes often become inhibitory agents as they alter the pH of surrounding environment.

Thus, phosphate or citrate buffers are added to maintain constant pH of the medium. Buffers are salts of weak acids or bases that keep the pH of the medium constant. Phosphate is a commonly used buffer and represents a good example of buffering by a weak acid ($H_2PO_4^-$) and its conjugate base (HPO_4^{2-}) in the following manner.

$$H^+ + H^+PO_4^{-2} \rightarrow H_2PO_4^-$$

$$OH^- + H_2PO_4 \rightarrow HPO_4^{-2} + HOH$$

Oxygen Requirements

The atmosphere of earth contains about 20% (v/v) of oxygen. Microorganisms capable of growing in the presence of atmospheric oxygen are called aerobes whereas those that grow in the absence of atmospheric oxygen are called as anaerobes.

The micro-organisms that are completely dependent on atmospheric oxygen for growth are called obligate aerobes whereas those that do not require oxygen for growth but grow well in its presence are called as facultative anaerobes.

Aerotolerants (e.g. Enterococcus faecalis) ignore O_2 and can grow in its presence or absence. In contrast, obligate anaerobes (e.g., Bacteroids, Clostridium pasteurianum, Furobacterium) do not tolerate the presence of oxygen at all and ultimately die. Few microorganisms (e.g., Campylobacter) require oxygen at very low level (2-10%) of concentration and are called as microaerophiles. The latter are damaged by the normal atmospheric level of oxygen (20%).

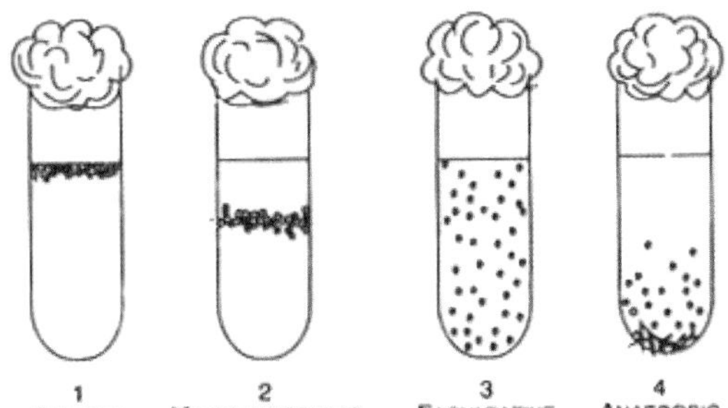

Relationship of oxygen and growth (shake tubes method)

These varying relationships between microbes (especially the bacteria) and O_2 appear due to different factors such as protein-inactivation and the effect of toxic oxygen-derivatives. Bacterial enzymes can be inactivated when interact with oxygen. Nitrogen-fixing enzyme introgenese is very sensitive to oxygen and represents a good example of interaction between enzyme and oxygen.

During metabolic process, flavoprotein reduces oxygen to form hydrogen peroxide (H_2O_2), superoxide radical (O_2^-) and hydroxy radical (OH^-). These compounds are highly toxic and being powerful oxidizing agent, can cause destruction of cellular macromolecules.

$O_2 + e^- \rightarrow O_2$ (superoxide radical)

$O_2^- + e^- + 2H^+ \rightarrow H_2O_2$ (hydrogen peroxide)

$H_2O_2 + e^- + H^+ \rightarrow H_2O + OH^-$ (hydroxy radical)

In order to survive, therefore, bacteria must be able to protect it from oxidising agents. All aerobes and facultative anaerobes contain two enzymes namely superoxide dismutase and catalase. These enzymes protect microbes against lethal effects of oxygen products.

The oxidizing property of superoxide is nullified by superoxide dismutase as it converts superoxide into oxygen and hydrogen peroxide. The enzyme catalase decomposes hydrogen peroxide into oxygen and water. The aero-tolerant bacteria like Lactic acid bacteria possess enzyme peroxidases instead of catalase to decompose the accumulated hydrogen peroxide.

$$2O_2^- + 2H^+ \xrightarrow[\text{dismutase}]{\text{superoxide}} O_2 + H_2O_2$$

$$2H_2O_2 \xrightarrow{\text{catalase}} 2H_2O + O_2$$

$$H_2O_2 + NADH + H^+ \xrightarrow{\text{Peroxidase}} 2H_2O + NAD^+$$

Since all obligate anaerobes lack these enzymes or have them in very low concentration and, therefore, are susceptible to oxygen.

Pressure

Normal life of microorganisms on land or on the water surface is always subjected to a pressure of 1 atmosphere. But, they are many microbes that survive in extremes of hydrostatic pressure in deep sea. Others are there that not only survive rather grow more rapidly at high pressures (e.g., Protobacterium, Colwellia, Shewanella) and are called barophilic.

Some archaebacteria are thermobarophiles (e.g., Pyrococcus spp., Methanococcus jannaschii). However, a barophile has been recovered from the depth about 10,500 m in sea near Philippines and has been found unable to grow at 2°C temp, and below about 400-500 atmospheric pressure.

Radiation

Some electromagnetic radiations, particularly the ionizing radiation (e.g., X-rays, gamma rays) are very harmful to microbial growth. Low levels of these radiations may cause mutations and may indirectly result in death whereas high levels may directly cause death of the microbes.

Ionizing radiation, however, destroys ring-structures, breaks hydrogen bonds, oxidizes double bonds and polymerizes certain molecules. Ultraviolet (UV) radiation is lethal to all categories of microbial life due to its short wavelength and high-energy; the most lethal UV radiation has a wavelength of 260 nm.

Ultraviolet radiation primarily forms thymine dimers in DNA to cause damage. Two adjacent thymines in a DNA strand join each other covalently and inhibit DNA replication and function. Microbial photosynthetic pigments (chlorophyll, bacteriochlorophyll, cytochromes and flavins), sometimes, absorb light energy, become excited or activated, and act as photosensitizers.

The excited photosensitizer (P) transfers its energy to oxygen which then results in singlet oxygen ($^1O_2$). The latter is very reactive and powerful oxidizing agent and quickly destroys a cell. The singlet oxygen is probably the main weapon employed by phagocytes to destroy engulfed bacteria.

Food Spoilage

Food spoilage can be defined as a disagreeable change in a food's normal state. Such changes can be detected by smell, taste, touch, or sight. These changes are due to a number of reasons -- air and oxygen, moisture, light, microbial growth, and temperature.

Air and Oxygen

One important cause of food spoilage and spoilage is air and oxygen. Because air is colorless, odorless, and tasteless, it is often taken for granted and sometimes forgotten as a means to cause food to spoil.

Air consists of 78% nitrogen, 21% oxygen, and a 1% mixture of other gases. While oxygen is essential for life, it can have deteriorative effects on fats, food colors, vitamins, flavors, and other food constituents. Basically, oxygen can cause food spoilage in several ways. It can provide conditions that will enhance the growth of microorganisms; it can cause damage to foods with the help of enzymes; and it can cause oxidation.

Microorganisms: Oxygen can provide conditions that enhance the growth of microorganisms. Some bacteria require oxygen for growth (aerobes) while others can grow only in the absence of oxygen (anaerobes). Many bacteria can grow under either condition and are called facultative anaerobes. Molds and most yeast that cause food to spoil require oxygen to grow. They can often be found growing on the surface of foods when air is present.

Enzymes: Certain enzymes that are naturally present in food are known as oxidizing enzymes. These enzymes catalyze (speed up) chemical reactions between oxygen and food components, and this leads to food spoilage. Although there are many oxidizing enzymes, two that can cause darkening in diced and sliced vegetables are catalase and peroxidase. The browning of vegetables caused by these enzymes is often accompanied by off-flavors and odors. A simple heat treatment (blanching) is used to inactivate these enzymes.

Oxygen: Oxygen can also cause spoilage of foods spontaneously, by itself. Oxidative spoilage is the chief cause of quality loss in fats and fatty portions of foods. When lipids oxidize, short chain carbon compounds are formed; these compounds have very strong odors and flavors and are very undesirable and unacceptable. The off-odors resulting from this type of spoilage are sharp and acrid and have been described as linseed oil-like, tallowy, fishy or perfume-like.

Moisture

Water is one of the most common substances on earth. It is an essential component of all foods. The amount of water in a food (known as percent water) influences the appearance, texture, and flavor of the food. All living organisms as well as food contain water. Water makes up about 70% or more of the weight of most fresh (unprocessed) foods. Even "dry" foods like beans, flour and cereals contain some water. Fresh fruits and vegetables contain the most water -- between 90% and 95% water. The amount of water in some common foods is shown below.

Food	%	Food	%
Apples	84	Lettuce	96
Bread	32-37	Mayonnaise	15
Butter	16	Milk	88
Corn Flakes	4	Peaches	89
Green Beans	92	Peanut Butter	2
Flour	12	Raisins	18
Jams, Jellies, Preserves	29	Strawberries	90
Sugar	0.5		

Although the water content of a food is expressed as a percent, this number does not reflect how the water exists in the food. Water in food is classified according to its availability, or biological activity and is either "free" or "bound." Free water is not bound to any components in a food; it can be used for microbial growth and is also available for chemical reactions. Bound water is physically bound to large (molecules) components in the food. It is not available to microorganisms for their growth and it cannot participate in chemical reactions.

Water or moisture greatly affects the keeping qualities of food. Excessive moisture pickup can result in product spoilage and spoilage by:

- Microorganisms -- microorganisms need water to dissolve the food they use. Water allows the food to get into bacterial, yeast and mold cells where it is used for energy and growth. Water also allows waste products to escape from the cells.
- Chemical Reactions -- the moisture in food also functions to allow chemical reactions to occur between components in the product.

Food spoilage and spoilage can occur when there are slight changes in relative humidity. Moisture can condense on the surface of a product and this can result in many common food defects. The molding of grain, soggy cereals, and the caking and lumping of dry products like powders and cake mixes can result from excessive moisture. Other defects such as mottling, crystallization and stickiness have also been observed. Moisture condensing on the surface of a food can also provide an environment for bacteria and molds to grow and multiply. Physical defects such as cracking, splitting and crumbling occur when excessive moisture is lost from foods.

Water is controlled in foods by:

- Microorganisms -- microorganisms need water to dissolve the food they use. Water allows the food to get into bacterial, yeast and mold cells where it is used for energy and growth. Water also allows waste products to escape from the cells.
- Chemical Reactions -- the moisture in food also functions to allow chemical reactions to occur between components in the product.

Food spoilage and spoilage can occur when there are slight changes in relative humidity. Moisture can condense on the surface of a product and this can result in many common food defects. The molding of grain, soggy cereals, and the caking and lumping of dry products like powders and cake mixes can result from excessive moisture. Other defects such as mottling, crystallization and stickiness have also been observed. Moisture condensing on the surface of a food can also provide an environment for bacteria and molds to grow and multiply. Physical defects such as cracking, splitting and crumbling occur when excessive moisture is lost from foods.

Water is controlled in foods by:

- Drying (dehydration), concentration and evaporation -- by removing water to a certain level, deteriorative reactions can be reduced or prevented. Examples of products preserved by these drying techniques include dry milk, potato flakes, drink mixes, evaporated milk, and orange juice concentrate.
- Freezing - the freezing of foods changes water from liquid to solid form and renders it unavailable to microorganisms and chemical reactions.
- Food Additives -- salt and sugar are used in many products to bind water and thereby making it less available for microbial growth and biochemical reactions. Jams, jellies, and cured hams are examples.

Light

Almost all foods are exposed to light from natural and/or artificial sources. The exposure of foods to light can result in the photodegradation (or spoilage) of these products. This photodegradation usually occurs in food constituents, such as pigments, fats, proteins, and vitamins and results in discoloration, off-flavor development, and vitamin losses.

The light that is absorbed by the food can cause deteriorative reactions of the food constituents. In most solid foods, the light only penetrates the outer layer of the product and photodegradation occurs in this surface layer. Discoloration on the surface of foods can certainly affect consumer acceptance of these products.

In liquid foods, light penetration can be greater and with mixing of the products due to agitation, larger portions of food constituents may be deteriorated. The light sensitivity of a food depends on many factors including the: light source strength and type of light that it emits; distance of the light source form the food; length of exposure optical properties of the packaging materials; oxygen concentration of the food; and the temperature.

Microbial Growth

Microorganisms, specifically bacteria, molds, and yeasts, can cause food to spoil. For example, microorganisms that break down fats in unsalted butter can cause it to become rancid. Bacteria that breaks down protein in meat (proteolytic bacteria) and results in foul odors. Also, if milk is kept too long or at improper temperatures, it will sour. Bacteria that have survived pasteurization grow in the milk and produce acid from the lactose that is naturally present in the milk.

The general sources of food spoilage microorganisms are the air, soil, sewage, and animal wastes. Some microorganisms that are naturally present on the surface of foods grown in the ground can also cause food spoilage. Bacteria from the animal's internal organs, skin, and/or feet can contaminate meat and fish. Meat is rapidly contaminated when it is ground for hamburger or sausage because the bacteria normally present on the outside of the meat move into the chopped meat where there are many air pockets and rich supply of moisture. Fish tissues are contaminated more readily than meat because they are more delicate and so more easily penetrated.

Canned foods undergo a sterilization process to make them shelf-stable. If canned foods are not properly processed, food spoilage (or food safety concerns) may occur. Swollen cans usually contain gas produced by the bacteria Clostridium. Sour spoilage without gas is commonly due to Bacillus. This type of spoilage is called flat-sour spoilage. Lactobacilli are responsible for acid spoilage when they break down the carbohydrates in foods and produce detectable amounts of acid.

Temperature

When temperatures are not properly controlled, food can spoil. For example, for every 18°F rise in temperature within the moderate temperature range where most food is handled (50°F to 100°F), the rate of chemical reaction is approximately doubled. As a result, excessive heat will increase the rate of natural food enzyme reactions and the reactions of other food constituents. As a result, protein will breakdown or denature; emulsions will break; some vitamins will be destroyed; moisture will be lost and foods will dry out; and the color, flavor and odor of some products may be affected.

Exposing foods to uncontrolled cold temperatures will also cause physical spoilage. Fruits and vegetables that accidentally freeze and thaw have their texture and appearance affected. Skins and surfaces of these products will often crack, leaving them more susceptible to microbial contamination. Some foods that become frozen may also be adversely affected. For example, if mayonnaise freezes, the emulsion will break and the components will separate.

Spoilage is caused in many foods by temperatures that are not extreme. Cold damage of several fruits and vegetables can occur at common refrigerator temperatures (35-40°F). Defects in produce exposed to cold temperatures include the development of off-colors, surface pitting and a variety of decays. Uncut, fresh fruits and vegetables such as bananas, lemons, squash and tomatoes are products that should be held at temperatures no colder than 50°F for best quality.

Meat Spoilage

Meat is a nutritious, protein-rich food which is highly perishable and has a short shelf-life unless preservation methods are used. It is the first-choice source of animal protein for many people all over the world. Consumption of meat is continuously increasing worldwide. Although consumption of meat is increasing world wide the significant portion of meat and meat products are spoiled every year. The significant portion of this loss is due to microbial spoilage.

Causes of Meat Spoilage

Pre-Slaughter Handling Effects on Meat Quality

Handling during the pre-slaughter period (that period after the animal is last removed from feed) is thought to have an effect on beef quality mainly through an effect on the ultimate pH of the beef by means of glycogen depleting events during this period. The different glycogen depleting events have an additive effect so that an animal that started the pre-slaughter period with a muscle glycogen at only 1% would have a muscle glycogen of 0.4% following a 0.6% glycogen depleting event, while an animal that started with 1.8% glycogen would have a muscle glycogen level of 1.2% following the same event.

This means that the same event has no effect on the ultimate pH of the latter animal but results in the former animal having a significantly elevated ultimate PH. If however, the latter animal was exposed to an additional 0.8% glycogen depleting event following the initial event, then it would have only 0.4% of glycogen at slaughter and would also have a significantly elevated ultimate PH. For most purposes an elevated ultimate PH results in beef of lower overall quality although some of the individual quality characteristics may be improved for certain purposes. Relative to beef with a normal rested ultimate pH of about 5.5, beef with an elevated ultimate PH will be characterized by:- A darker color, especially when the ultimate PH is above about 6.2, and this effect has given rise to the term "dark-cutting beef", A blander and usually less acceptable flavor. If foreign flavours are present, these become more apparent when the basic beef flavor is weaker, a shorter shelf life because some spoilage bacteria grow better at the higher PH and because fewer bacteria are needed to produce off-flavors when the pH is higher, A higher water-holding capacity, which may be an advantage for beef to be used for manufacturing purposes, Being tougher if the ultimate pH is in the intermediate range (PH of 5.8to 6.2), but as tender or more tender than an animal with a PH of around 5.5, if a high PH exists (PH of 6.5 to 7.0).

Overall, the aim during the pre-slaughter period, apart from maximizing the welfare of the animals should be to minimize the likelihood of elevated ultimate PH values. This can be done by a combination of having cattle accustomed to being handled, by avoiding excessively long holding times, by minimizing potentially stressful situations (strange noises, sights and mixing with strange animals), and by feeding animals well prior to this period so that muscle glycogen levels are not depleted or marginal. Bring cattle off pasture 4 hours or more before transport so they are easier to handle. However, minimize the fasting period to less than 8 hours to avoid carcass weight losses. Alternatively, provide hay or silage to avoid weight loss and improve behavior.

The glycogen content of animal muscles is reduced when the animal is exposed to pre-slaughter stress which changes the pH of the meat, to higher or lower levels, depending on the production level of lactic acid. Lactic acid is produced due to the breakdown of glycogen content of animal muscles via an anaerobic glycolytic pathway as shown in the following figure;

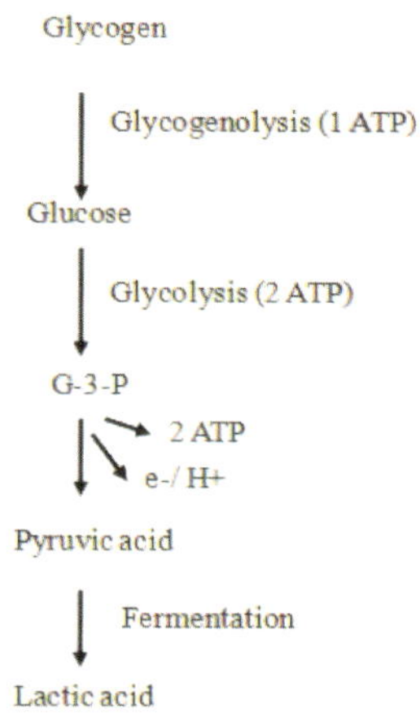

Anaerobic glycolytic pathway

Higher levels of pH (6.4-6.8) result in Dark, Firm and Dry (DFD) meat. Long term stress causes DFD meat which has a shorter shelf life. (Chambers and Grandin, 2001). Severe short term stress results in a Pale, Soft and Exudative (PSE) meat. PSE meat has a pH lower than normal ultimate value of 6.2 which is responsible for the breakdown of proteins, providing a favorable medium for the growth of bacteria (Miller, 2002). The following figures show the texture and color of DFD, PSE and normal meat.

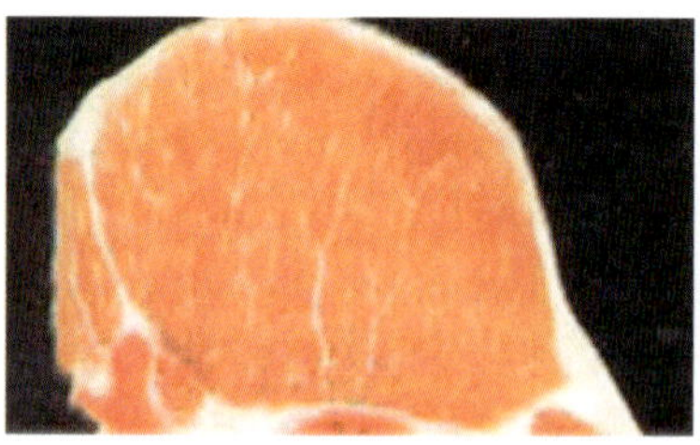
(A) Normal meat.

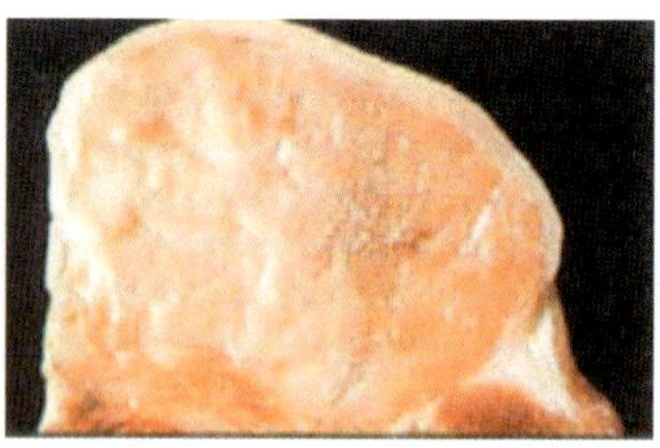
(B) Pale Soft and Exudative (PSE) meat.

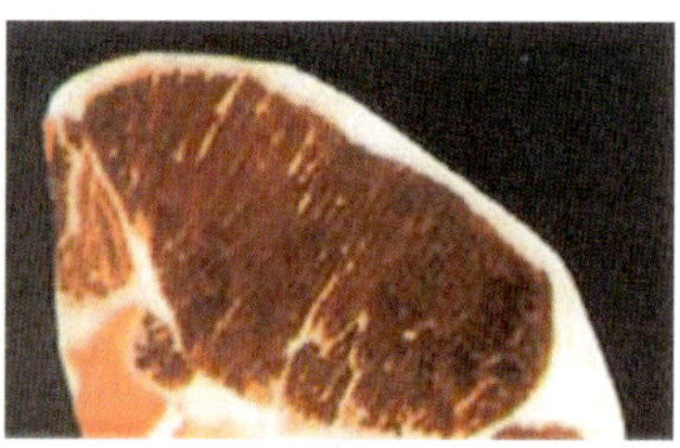
(C) Dark Firm and Dry (DFD) meat.

Nutritional Effects on Meat Quality

Nutritional effects can be divided into the effects of the level of nutrition and the effects of the composition of the diet. In both cases there will often be an effect on growth rate, so that in order to compare two animals at the same weight they will be at different ages. It is quite possible that even at the same weight they will have a different level of fatness so it is often difficult to distinguish between indirect effects of a nutritional treatment through differences in age and/or fatness, and direct effects (e.g. of specific nutrients).Not surprisingly, many trials have been conducted in this area and it is not possible to do justice to all the results that have been published, because for just about every significant effect that has been reported it is possible to find another similar trial where the effect did not occur.

The limited number of effects that have arisen with reasonable consistency, that would lead to beef from one animal being superior in one or more ways to that from another because of the amount of feed it received or because of the nature of that feed: 1) If one animal is fed in such a way that growth rates for the few months before slaughter are higher than those for another, some evidence suggests that the beef from the former animal is likely to be more tender. If the higher growth rates are the result of a high-grain ration (as is usual on a feedlot) the flavor of the beef will differ as well, but whether or not one is preferred will depend on the consumer involved. The greater tenderness may be due to several factors, but higher levels of marbling, and slightly more soluble collagen have both been suggested as possiblecauses.2) The effects of specific nutrients on beef flavour are less common than for meat from monogastric animals such as the pig because of the neutralizing effects of rumen metabolism, but some flavor effects have been reported, such as that associated with grain feeding.

Transportation

Animals are reared on farms which may be situated far away from other farms, markets and slaughter plants. Therefore, they have to be transported over some distances to such locations either for better and less expensive feed, sale and slaughter. Transportation begins with loading and ends with unloading. Both thought to be done in a gentle manner and under a quieter environmental condition. During transportation, animals are exposed to environmental stress such as heat, cold, humidity, noise and overcrowding. Over speeding, sudden stops, rapid acceleration and long journey times without appropriate rest should be avoided as these will increase the spate of carcass and meat quality defects. Through the processes of rounding up, loading, transporting and off-loading cattle at the abattoir, there are animal welfare compromises. Fear of either the handling procedures and novelty of the environment makes the animal feel pain.Transportation conditions appear to be by far the most common reported by most workers. This indicates that transportation is a critical point in the life of an animal prior to slaughter. Transportation is strange and life threatening to animals. During transportation, animals can do little to help themselves when they are in danger or discomfort due to confinement. In farms, markets and lair age, animals have much freedom of movement and could avoid certain uncomfortable conditions. Other conditions such as starvation, dehydration, injury, suffocation, heat or cold stress, overcrowding, death and many more do prevail in transportation.

Marketing

Animals are either sold at the markets or sent directly from farms to abattoirs for slaughter. In the markets they may be kept in groups and in open pens which may expose them directly to the sun or cold. They may encounter stresses such as noise, unfamiliar environment and social regrouping. They can also be starved or dehydrated if feed and water is redrawn longer than recommended. Various degrees of bruising can occur on their skin at the market depending on the way they are handled. Stress experienced during routine handling and restraint procedures reduces productivity and Apart from causing poor meat quality and economic losses, stress has negative implications on animal welfare.

Lairaging

Animals are held temporarily in the lairage prior to slaughter. It serves as a collection point for different animals just before slaughter. The lairage is also used to provide animals with some amount of recovery from stress during transport. Animal's movement should be observed for any possible injury or infection. They should also be given adequate amount of suitable feed and water if they will be kept in the lairage longer than expected. Although lairaging is to enable animals to rest and to recover from transportation stress, it can be a major source of meat quality problems. Animals may suffer from different degree of bruising and injury as a result of fighting or overcrowding. Lairage can also act as reservoirs of infection by pathogenic bacteria and there is evidence that longer holding times increase the risk of carcass contamination.

Careless and improper handling of animals in the lairage such as the use of electrical goads, abstracting the movement of animals through race, beating and firm grip of the coat with the hand, exposure of animals to microbial contaminations will adversely have effect on the carcass and meat quality. Conditions in the lairage therefore need to be conducive to prevent infections and the animals from being stressed further after vigorous transportation but to recover from transportation stress. During lairage, animals are subjected to ante-mortem inspection with the following aims: a) demanding and checking the health and vaccine certificates of the cattle; b) identifying the hygienic and health status to aid, with informative data, the post-mortem inspection tasks; c) identifying and isolating sick or suspicious animals before slaughter, as well as late pregnancy and recently calved cows; d) verifying the hygiene conditions of the pens and annexes.

Effects of Stunning and Slaughter on Meat Quality

Stunning methods can have adverse effects on carcass and meat quality and cause downgrading. These could be visual effects such as bruising and hemorrhages, pelt burn in sheep, bone fractures, color changes caused by DFD as well as those manifested in eating quality such as toughness. During electrical stunning blood pressure changes, muscle spasms and convulsions can cause ruptures and hemorrhages in vessels and muscle as well as fractures. Post-mortem metabolism can be influenced by indirect stimulation by nerves. Broken vertebrae can occur when stunned with head to back electrode positioning if the voltage and the current is too high.

Although hemorrhages can be induced by stunning and killing, the underlying mechanism is thought to be multi factorial. Histological studies on structures where hemorrhages occurred showed that blood leaking out of vessels is determined by the type of surrounding tissue and also the amount of blood leaving the circulation. Some hemorrhages were associated with hyper contracted and disrupted muscle fibers, indicating that they were caused by severe muscular strain. Many hemorrhages were found near venules or veins where rupture was observed, not in arterial vessels. This indicates that venous blood pressure increase can cause rupture of venules and small veins. In order to reduce petechial hemorrhages and bruising following can be considered: 1) Shorten stunning to sticking interval so that blood leakage through ruptured vessels is reduced. 2) Captive bolt stunning may be preferable to electrical stunning if blood splash is a problem as muscle spasms are less pronounced after captive bolt. 3) Electrical stunning currents are applied in a continuous and uninterrupted manner. 4) In lambs electrical stunning with cardiac arrest may reduce blood pressure and blood splash.

Post Slaughter Handling of Meat

There are three main mechanisms for meat and meat products spoilage after slaughtering and during processing and storage: microbial spoilage, lipid oxidation and autolysis enzymatic spoilage.

- Microbial Spoilage

Meat and meat products provide excellent growth media for a variety of micro flora (bacteria, yeasts and molds) some of which are pathogens. The intestinal tract and the skin of the animal are the main sources of these microorganisms. The composition of micro flora in meat depends on various factors: (a) preslaughter husbandry practices (free range Vs intensive rearing), (b) age of the animal at the time of slaughtering, (c) handling during slaughtering, evisceration and processing, (d) temperature controls during slaughtering, processing and distribution (e) preservation methods, (f) type of packaging and (g) handling and storage by consumer.

The most common genera of bacteria found in meat before spoilage is Staphylococcus, Bacillus, Campylobacter, Clostridium, Listeria, Salmonella etc. Mold species found in meat include Cladosporium, Geotrichum, Penicillium and Mucor while yeasts species include Candida spp and Cryptococcus sp. The storage conditions affect the type of microbes found in meat and meat products. The favorable pH for the growth of spoilage bacteria for meat is in the range of 5.5-7.0. Slime formation, structural components degradation, off odors and appearance change were found in meat as a result of microbial growth within this pH range.

- Lipid Oxidation

Autoxidation of lipids is natural processes which affect fatty acids and lead to oxidative deterioration of meat and off-flavours development. After slaughtering of animals, the fatty acids in Tissues undergo oxidation when the blood circulation stops and metabolic processes are blocked. Oxidation of lipids in meat depends on several factors including: fatty acid composition, the level of the antioxidant vitamin E and prooxidants such as the free iron presence in muscles. Poly saturated fatty acids are more susceptible to lipid oxidation. Hydro peroxides are produced due to the lipid oxidation of highly unsaturated fatty acid fractions of membrane phospholipids, which are susceptible to further oxidation/ decomposition. Their breakage produce oxygenated compounds such as aldehydes and ketones.

These secondary products can cause loss of color and nutritive value due to sever effects on lipids, pigments, proteins, carbohydrates and vitamins.

In meat, lipid hydrolysis can take place enzymatic ally or non-enzymatic ally. The enzymatic hydrolysis of fats is termed lipolysis or fat deterioration and is governed by specific enzymes such as lipases, esterase and phospholipase. Lipolytic enzymes could either be endogenous of the food product (such as milk) or derived from psychrotrophic microorganisms. Lipases enzymes are present in the skin, blood and tissue of animals. During lipolysis, lipases split the glycerides forming free fatty acids which are responsible for common off-flavour, frequently referred to as rancidity. The main enzymes involved in meat lipid hydrolysis are phospholipase A1 and phospholipase A2. The non-enzymatic hydrolysis is caused by heme proteins such as hemoglobin, myoglobin and cytochrome which are susceptible to oxidation and produce hydroperoxides.

- Autolytic Spoilage

Enzymatic actions are natural process in the muscle cells of the animals after they have been slaughtered and are the leading cause of meat deterioration. The enzymes have the ability to combine chemically with other organic compounds and work as catalysts for chemical reactions that finally end up in meat self deterioration. In the autolysis process, the complex compounds (carbohydrates, fats and protein) of the tissues are broken down into simpler ones resulting in softening and greenish discoloration of the meat. These autolysis changes include proteolysis and fat hydrolysis which are prerequisite for microbial decomposition. Excessive autolysis is termed "souring. Postmortem breakdown of polypeptides are the result of tissue proteases and is responsible for flavour and is textural changes in meat. The enzymes calpains, cathepsins and amino peptidases are found to be responsible for the post mortem autolysis of meat through digestion of thc z- line proteins of the myofibril. Among these enzymes, calpains has been described as a preliminary contributor to the proteolytic tenderization process of meat. Proteolytic enzymes are active at low temperatures (5°C) which lead to deterioration of meat quality due to growth of microbes and biogenic amines production.

Preservation of Meat

Meat preservation became necessary for transporting meat for long distances without spoiling of texture, colour and nutritional value after the development and rapid growth of super markets. The aims of preservation methods are to inhibit the microbial spoilage and to minimize the oxidation and enzymatic spoilage. Traditional methods of meat preservation such as drying, smoking, brining and canning have been replaced by new preservation techniques such as chemical, biopreservative and nonthermal techniques. Current meat preservation methods are broadly categorized into three methods. Controlling temperature, controlling water activity and use of chemical or biopreservatives and combination of these preservation techniques can be used to diminish the process of spoilage.

Low Temperature Methods:

The basic aim of cooling techniques is to slow or limit the spoilage rate as temperature below the optimal range can inhibit the microbial growth. Low temperature methods of storage are used in three levels: chilling, freezing and super chilling. All these levels help to inhibit or completely stop bacterial growth.

- Chilling

Chilling is employed at slaughtering plants immediately after slaughtering and during transport and storage. It is necessary to reduce the temperature of carcass immediately after evisceration to 4°C within 4 h of slaughtering. Chilling is critical for meat hygiene, safety, shelf life, appearance and nutritional quality. Chilling also helps to prevent denaturing of proteins which may lead to bacterial attack as they are more susceptible to denaturated protein than native protein. On the other hand, cold-shortening and toughening may result from ultra-rapid chilling of pre-rigour meat. It is employed by two methods: immersion chilling, in which the product is immersed in chilled (0- 4°C) water and air chilling in which the carcasses are misted with water in a room with circulating chilled air. Carcass surface temperature is reduced at faster rate by air chilling which improves carcass drying and minimizes microbial spoilage. The microbial quality of the air-chilled product is better than that of a water-chilled product. The air-chilled carcasses lost 0.68% of their post slaughter weight in storage prior to cutting but lost no more during cutting or post cutting storage. On other hand water chilled carcasses absorbed 11.7% moisture in thc chillers, of which 4.72% is lost within 24 h of intact carcass storage, 0.98% is lost during cutting and 2.10% is lost during storage resulting in 3.9% net water retention.

- Freezing

Freezing is an excellent method of keeping the original characteristics of fresh meat. Meat contains about 50-75% by weight water, depending on the species, and the process of freezing converts most of water into ice. Meat freezing phenomenon is fast and almost 75% of tissue fluid freezes at -5°C. The freezing rate is increased with decreases in temperature, almost 98% of water freezes at -20°C and complete crystal formation occurs at - 65°C. However, more than 10% of muscle bound water (chemically bound to specific sites such as carbonyl and amino group of proteins and hydrogen bonding) will not freeze. Freezing rate (slow and fast) affects the quality of frozen meat significantly. Fast freezing produce better quality meat than slow freezing.

The preservation capacity of frozen meat is limited because the physical, chemical or biochemical reactions that take place in animal tissues after slaughtering do not stop absolutely after cold treatment. Microbial growth stops at -12°C and total inhibition of the cellular metabolism in animal tissues occurs below - 18°C. Complete quality changes of meat can be prevented at a temperature of-55°C. However, enzymatic reactions, oxidative rancidity and ice crystallization will still play an important part in spoilage. During freezing, about 60% of the viable microbial population dies but the remaining population gradually increases during frozen storage.

Pathogenic microorganisms are commonly isolated from thawed frozen meat. The shelf-life of vacuum-packaged fresh beef is approximately 35-45 days; longer shelf-life of 70-80 days is possible with refrigeration of 0-2.3°C. Vacuum packaged frozen, whole-muscle beef has a recommended shelf-life of 12 months. Low (-18°C) and constant storage temperature substantially increase the shelf life of meat. Shelf life of red meat stored at 15°C-30°C normally ranges from 6 months to 24 months.

- Super Chilling

Super chilling is a different concept than refrigeration and freezing and it has the potential to reduce storage and transport costs. Super-chilling refers to the temperature zone below its initial freezing point (1-2°C) but where ice crystals are not generated. In this process, instead of adding external ice to the food product, part of the internal water is frozen and works as a refrigeration reservoir, ensuring its refrigeration during distribution and transportation. The main advantage of this method of preservation over traditional methods is that it increases the shelf life of meat for up to 4 times. Although most microbial activities are stopped or inhibited, chemical and physical changes may progress and in some cases are even accelerate.

Controlled Water Activity Methods

Microbiological safety of food is directly influenced by the water activity (aw). The term water activity (aw) refers to water which support the growth of microorganisms. It represents the ratio of the water vapour pressure of the food to the water vapour pressure of pure water under the same conditions. Water activity in meat products is equivalent to the relative humidity of air in equilibrium with the product. Most fresh meats, fruits and vegetables fall into moist food category, have a water activity more than 0.85 and require refrigeration or another barrier to control the growth of pathogens. Each microorganism has minimum, optimum and maximum water activities. Micro-organisms generally grow best between aW values of 0.980-0.995 and growth ceases at aW<0.900. Yeasts and molds can grow at a low aW of 0.6. However, growth of pathogens is prevented at aW of 0.85. In processed and cured meats, the growth of gram negative bacteria (that can tolerate an aW of 0.94-0.97) can be suppressed with reducing water activity.

Table: The minimum water activity for growth of the most common microorganisms associated with dried meat products (USDA, 2005).

Microorganisms	Water activity
Campylobacter	0.98
Pseudomonas	0.97
Clostridium botulinum(non proteolytic)	0.96
Clostridium botulinum(proteolytic)	0.93
Salmonella	0.94
Clostridium perfringens	0.93
Escherichia coli 0157:H7	0.95
Listeria monocytogens	0.92
Staphylococcus aeurus(anaerobic)	0.90
Staphylococcus aeurus	0.86
Aspergilus flavus	0.80

Water activity in meat is controlled by drying, refrigeration, adding chemicals or a combination of these methods. Sodium chloride and sugar are used to control water activity as free water binds up in their presence which results in an osmotic imbalance and finally inhibition of cell growth.

Sodium Chloride

Sodium chloride (NaCl) in growth media or foods can be a source of osmotic stress by decreasing water activity. Salt-sensitive microorganisms, such as Pseudomonas spp. do not grow in meat when the water activity (aW) is reduced from 0.99 to 0.97 with the addition of 4% sodium chloride. However, salt tolerant microorganisms such as lactic acid bacteria and yeasts could grow at that level of water activity.

Sugars

Sugars have the capabilities to bind with moisture and reduce water activity in foods. Dextrose, sucrose, brown sugar, corn syrup, lactose, honey, molasses and starches are generally used in dried meat processing as a source of sugars or carbohydrates to enhance flavor, reduce harshness of salt and lower water activity.

Chemical Methods

Energy intensive freezing operations are the greatest way to preserve carcass, meat and meat products for a longer time which inhibits bacterial growth, but not the psychrophiles and the spores. Most of these survive freezing and grow during thawing. Traditional methods for preservation of meat by salting and picking are well accepted procedures. Other chemicals have been used as food additives for preservation of meat but every country has drawn its rules and regulations and established limits for the purpose of prevention of harmful effects to human.

Antimicrobial preservatives are substances which are used to extend the shelf life of meat by reducing microbial proliferation during slaughtering, transportation, processing and storage. Growth of bacteria and spoilage of meat is depending on the species of bacteria, nutrients availability, pH, temperature, moisture and gaseous atmosphere. Antimicrobial compounds added during processing should not be used as a substitute for poor processing conditions or to cover up an already spoiled product. They offer a good protection for meat in combination with refrigeration. Common antimicrobial compounds include: Chlorides, Nitrites, Sulfides and Organic acids.

Sodium Chloride

Sodium chloride has a long history of use in food preservation in sufficiently high Concentrations. It inhibits microbial growth by increasing osmotic pressure as well as decreasing the water activity in the microenvironment. The combination of sodium chloride with other antimicrobial agents may have an impact on the overall inhibitory effect. A combination of NaCl and sodium lactate is more effective than lactates alone in delaying the onset of meat spoilage and its effects on its color and fat stability. The use of sodium chloride in combination with sodium lactate reduces the microbial growth, maintain the chemical quality and extend the shelf life of ground beef during refrigerated storage.

Nitrites

The nitrites used in meat preservation industry are always in the form of salts such as sodium nitrite or potassium nitrite. Nitrites provide stabilized red meat color, cured meat flavor and rancidity retardation. Nitrite salts are effective in controlling color, lipid oxidation and odor in addition to controlling the anaerobic bacteria.

Nitrites affect the growth of microorganisms in food through several reactions including: (a) reacting with alphaamino groups of the amino acids at low pH levels, (b) blocking sulfhydryl groups which interferes with sulphur nutrition of the organism, (c) reacting with iron-containing compounds which restricts the use of iron by bacteria, and (d) interfering with membrane permeability which limits the transport across cells.

Sulphites

As antimicrobial agent, sodium sulfite is efficient against aerobic Gram-negative bacilli, molds and yeasts in meat and meat products. The antimicrobial activity is the result of the un dissociated sulfurous acid which enters the cell and reacts with thiol groups of proteins, enzymes and cofactors. Yeast cells are attacked by sulfite because sulfite reacts with cellular Adenosine Triphosphate (ATP) and blocks the cystine disulphide linkages.

Lactic Acid

Lactic acid is shown antimicrobial activities against many pathogenic organisms such as Clostridium botulinum because of its abilities to reduce pH level, exert feedback inhibition and interferes with proton transfer across cell membranes.The salt of lactic acid (lactate) is used in the meat industry as an antimicrobial agent. The use of lactic acid bacteria as inoculums is a newly developed approach for food preservation. Lactic acid bacteria are effective in inhabiting undesirable microorganisms in food by producing a wide range of substances (such as lactic acid, acetic acids and hydrogen peroxide,) which inhabit the growth of other microorganisms.

Sorbic Acid

Sorbic acid (2, 4-hexadienoic) and its salts are widely used throughout the world as meat preservatives for inhibiting bacteria and fungi. A concentration of 0.3% sorbates in food is high enough to inhibit the microorganisms. The sorbic acid has an inhibitory mechanism via depression of internal pH. Sorbates can interfere with the bacterial spore germination, inhibit the activity of several enzyme systems and interfere with substrate and electron transport mechanisms.

Chemical Methods for Controlling Oxidative Spoilage

Freeze storage cannot prevent oxidative spoilage and microbial/enzymatic spoilage. Thus, chemical preservation methods are quite beneficial in combination with refrigeration in order to optimize stability, product quality while maintain freshness and nutritional value. Thorough understanding of lipid oxidation and its inhibition is necessary to prevent the development of rancidity, off flavour and discoloration in meat. Antioxidants can be classified as primary or long term antioxidants and secondary or processing antioxidants. Primary antioxidants include phenolic compounds while the secondary antioxidants include phosphates. Among the widely used lipid oxidation inhibitory additives in meat are: phenolic antioxidants (primary antioxidants) and phosphates (secondary antioxidants).

Phenolic Antioxidants

Derivatives of phenol such as Butylated Hydroxyanisole (BHA) and Proply Gallates

(PG) are referred to as synthetic phenolic antioxidants. Their use is extensive with the intention to delay, retard or prevent the negative effects of lipid per oxidation by scavenging chain-carrying peroxyl radicals or diminishing the formation of initiating lipid primary radicals (chain- breaking) and secondary radicals (preventive antioxidants).

Phosphates

Among the antioxidants in food additives, a phosphate is one of the first investigated for their potential antioxidant activities in meat products. A range of functionalities is provided by phosphates to enhance meat, poultry and sea food products. Functionalities of phosphate salts vary with the type of phosphate salt or combination of them. Phosphates critical functions include: (a) optimizing the water binding capacity of the muscle proteins by influencing pH, (b) interacting with muscle fibers for improved emulsification of fats, (c) maintaining the stability of the protein-fat-water system, (d) chelating divalent cations and retard rancidity which increase shelf life and (e) binding iron into the system and reducing oxidation.

Chemical Methods for Controlling Autolytic Enzymatic Spoilage

Autolysis is a term used to describe a series of postmortem chemical changes in the tissues of animals after death due to the presence of the enzymes (lipolytic, amylolytic and proteolytic) responsible for the metabolic process during the life of animals which are responsible for degradation of fats, carbohydrates and proteins after the death of animal. Lypolytic enzymes are responsible for the fat deterioration or lypolysis (oxidation) while

amylolytic enzymes are responsible for the change of glycogen to lactic acid. The enzymatic activities of calpain, cathepsins and aminopeptidases enzymes are affected by pH and temperature. The membrane of the lysosomes becomes leaky as the pH of the meat decreases post mortem and the enzymes are released. Moderate rates of postmortem pH decline (from 6.9-6.2after 3 h) allows for greater postmortem protein degradation and increased tenderization while rapid rates (from 6.9-5.8 in 3 h) and slow rates (from 6.9-6.6 in 3 h) produce less tender meat. Curing salts and acids have been used to inhibit the activity of such autolytic enzymes and prevent or slows degradation and spoilage.

Wine Fault

Wine faults are defined as off-aromas or tastes that are related to the fermentation or the winemaking process. Wine taints are defined as an odour, taste or aftertaste that originate from external or foreign sources.

There are many different, naturally occurring causes of wine taint, as well as several different types of wine taint. Each type of wine taint is identifiable through specific aromas and tastes.

Sulfides and Sulfur Compounds

Caused by a lack of amino acids in the fermenting wine, most commonly red wines, in small doses, sulfur adds complexity. Winemakers can avoid sulfur taint by adding Diammonium Phosphate during fermentation. Consumers can identify sulfur taint through the following smells:

- Boiled eggs
- Onion Garlic
- Burned rubber
- Tinned corn
- Cooked vegetables
- Oxidation

The most common wine fault especially in light bodied reds and white wines, oxidation occurs when there is a surplus of oxygen and catalyst in wine. Oxidization can even occur after a wine has been bottled. Consumers can identify oxidization through the following smells:

- Sherry
- Brown apple
- Paper

Cork Taint

Though the misconception is that cork taint is common, it actually isn't as common as it used to be. There are only a few main causes of cork taint, 2,4,6-trichloroanisole (TCA), Geosmin, Octenol and Methylisoborneol. Consumers can identify cork taint through the following smells:

- Must
- Earth
- Mushroom
- Compost

Heat Damage

Also known as cooked wine, this is caused when wine is exposed to intense heat that literally cooks the wine. Because heat damage can cause damage to the seal of a bottle, it is often combined with oxidization. Consumers can identify heat damage through the following characteristics: Jammy Sweet, but processed canned fruit.

UV Light Damage Also known as lightstrike, this is caused when wine gets too much sun or UV exposure. Most commonly found in delicate white wines, like Champagne, Viognier and Chenin Blanc. Consumers can identify it by the taste and smell of wet sweater. Mousy Taint Attributed to Brettanomyces, mousy taint can also be caused by the lactic acid bacteria in Malolactic fermentation. Not a volatile taint, it's difficult to identify through aroma alone, but it is apparent on the palate and in the back of the mouth in the form of mouse, mouse cage or mouth urine.

Secondary Fermentation or Refermentation

When there is residual yeasts left in the wine when it was bottled, it can cause an unintentional secondary fermentation in the bottle. This results in a slightly fizzy sensation in the mouth. When the fizz is intentional, the wine will be labeled as "frizzante" or "sparkling".

Differences between Flaws and Faults

In wine tasting, there is a big distinction made between what is considered a *flaw* and a *fault*. Wine flaws are minor attributes that depart from what are perceived as normal wine characteristics. These include excessive sulfur dioxide, volatile acidity, *Brettanomyces* or "Brett aromas" and diacetyl or buttery aromas. The amount to which these aromas or attributes become excessive is dependent on the particular tastes and recognition threshold of the wine taster. Generally, a wine exhibiting these qualities is still considered drinkable by most people. However, some flaws such as volatile acidity and *Brettanomyces* can be considered a fault when they are in such an excess that they overwhelm other components of the wine. Wine faults are generally major attributes that make a wine undrinkable to most wine tasters. Examples of wine faults include acetaldehyde (except when purposely induced in wines like Sherry and Rancio), ethyl acetate and cork taint.

Detecting Faults in Wine Tasting

The vast majority of wine faults are detected by the nose and the distinctive aromas that they give off. However, the presence of some wine faults can be detected by visual and taste perceptions. For example, premature oxidation can be noticed by the yellowing and browning of the wine's color. The sign of gas bubbles in wines that are not meant to be sparkling can be a sign of refermentation or malolactic fermentation happening in the bottle. Unusual breaks in the color of the wine could be a sign of excessive copper, iron or proteins that were not removed during fining or filtering. A wine with an unusual color for its variety or wine region could be a sign of excessive or insufficient maceration as well as poor temperature controls during fermentation. Tactile clues of potential wine faults include the burning, acidic taste associated with volatile acidity that can make a wine seem out of balance.

Wine fault	Characteristics
Acetaldehyde	Smell of roasted nuts or dried out straw. Commonly associated with Sherries where these aromas are considered acceptable
Amyl-acetate	Smell of "fake" candy banana flavoring
Brettanomyces	Smell of barnyards, fecal and gamey horse aromas
Cork taint	Smell of a damp basement, wet cardboard or newspapers and mushrooms
Butyric acid	Smell of rancid butter
Ethyl acetate	Smell of vinegar, paint thinner and nail polish remover
Hydrogen sulfide	Smell of rotten eggs or garlic that has gone bad
Iodine	Smell of moldy grapes
Lactic acid bacteria	Smell of sauerkraut
Mercaptans	Smell of burnt garlic or onion
Oxidation	Smell of cooked fruit and walnuts. Also detectable visually by premature browning or yellowing of the wine
Sorbic acid plus lactic acid bacteria	Smell of crushed geranium leaves
Sulfur dioxide	Smell of burnt matches. Can also come across as a pricking sensation in the nose.

Oxidation

The oxidation of wine is perhaps the most common of wine faults, as the presence of oxygen and a catalyst are the only requirements for the process to occur. Oxidation can occur throughout the winemaking process, and even after the wine has been bottled. Anthocyanins, catechins, epicatechins and other phenols present in wine are those most easily oxidised, which leads to a loss of colour, flavour and aroma - sometimes referred to as *flattening*. In most cases compounds such as sulfur dioxide or erythorbic acid are added to wine by winemakers, which protect the wine from oxidation and also bind with some of the oxidation products to reduce their organoleptic effect. Apart from phenolic oxidation, the ethanol present within wine can also be oxidised into other compounds responsible for flavour and aroma taints.

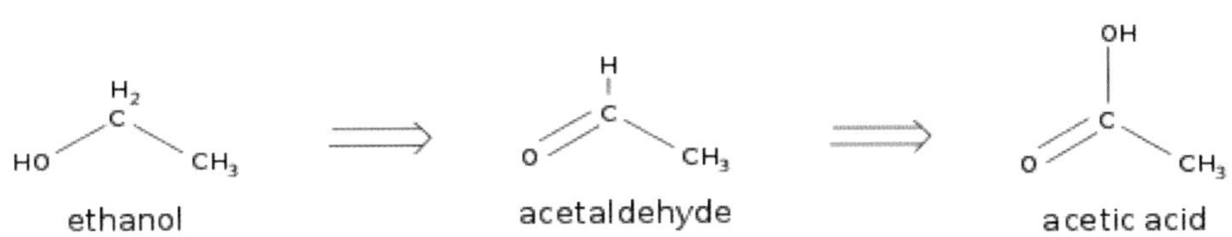

The oxidation of ethanol

Acetaldehyde

Acetaldehyde is an intermediate product of yeast fermentation; however, it is more commonly associated with ethanol oxidation catalysed by the enzyme ethanol dehydrogenase. Acetaldehyde production is also associated with the presence of surface film forming yeasts and bacteria, such as acetic acid bacteria, which form the compound by the decarboxylation of pyruvate. The sensory threshold for acetaldehyde is 100-125 mg/L. Beyond this level it imparts a *sherry* type character to the wine which can also be described as *green apple*, *sour* and *metallic*. Acetaldehyde intoxication is also implicated in hangovers.

Acetic Acid

Acetic acid in wine, often referred to as volatile acidity (VA) or *vinegar taint*, can be contributed by many wine spoilage yeasts and bacteria. This can be from either a by-product of fermentation, or due to the spoilage of finished wine. Acetic acid bacteria, such as those from the genera *Acetobacter* and *Gluconobacter* produce high levels of acetic acid. The sensory threshold for acetic acid in wine is >700 mg/L, with concentrations greater than 1.2-1.3 g/L becoming unpleasant.

There are different opinions as to what level of volatile acidity is appropriate for higher quality wine. Although too high a concentration is sure to leave an undesirable, 'vinegar' tasting wine, some wine's acetic acid levels are developed to create a more 'complex', desirable taste. The renowned 1947 Cheval Blanc is widely recognized to contain high levels of volatile acidity.

Ethyl acetate is formed in wine by the esterification of ethanol and acetic acid. Therefore, wines with high acetic acid levels are more likely to see ethyl acetate formation, but the compound does not contribute to the volatile acidity. It is a common microbial fault produced by wine spoilage yeasts, particularly *Pichia anomala* or *Kloeckera apiculata*. High levels of ethyl acetate are also produced by lactic acid bacteria and acetic acid bacteria.

Sulfur Compounds

Sulfur is used as an additive throughout the winemaking process, primarily to stop oxidation as mentioned above but also as antimicrobial agent. When managed properly in wine, its presence there is often undetected, however when used recklessly it can contribute to flavour and aroma taints which are very volatile and potent. Sulfur compounds typically have low sensory thresholds.

Sulfur Dioxide

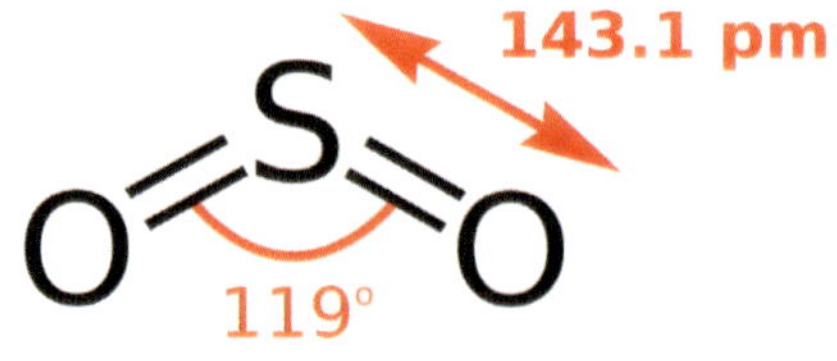

Sulfur dioxide is a common wine additive, used for its antioxidant and preservative properties. When its use is not managed well it can be overadded, with its perception in wine reminiscent of *matchsticks*, *burnt rubber*, or *mothballs*. Wines such as these are often termed *sulfitic*.

Hydrogen Sulfide

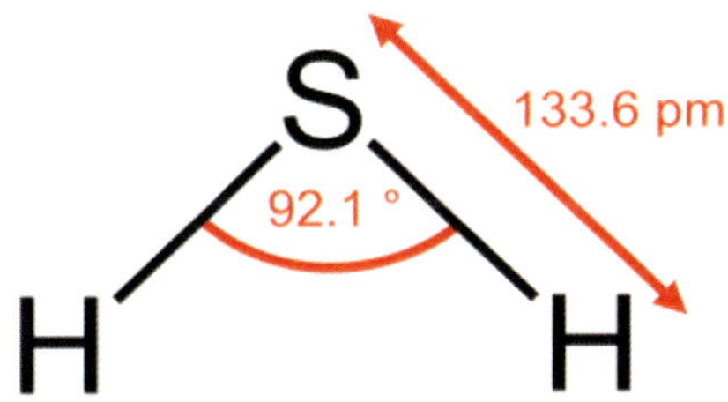

Hydrogen sulfide (H_2S) is generally thought to be a metabolic by-product of yeast fermentation in nitrogen limited environments. It is formed when yeast ferments via the sulfate reduction pathway. Fermenting wine is often supplemented with diammonium phosphate (DAP) as a nitrogen source to prevent H_2S formation. The sensory threshold for hydrogen sulfide is 8-10 μg/L, with levels above this imparting a distinct *rotten egg* aroma to the wine. Hydrogen sulfide can further react with wine compounds to form mercaptans and disulfides.

Mercaptans

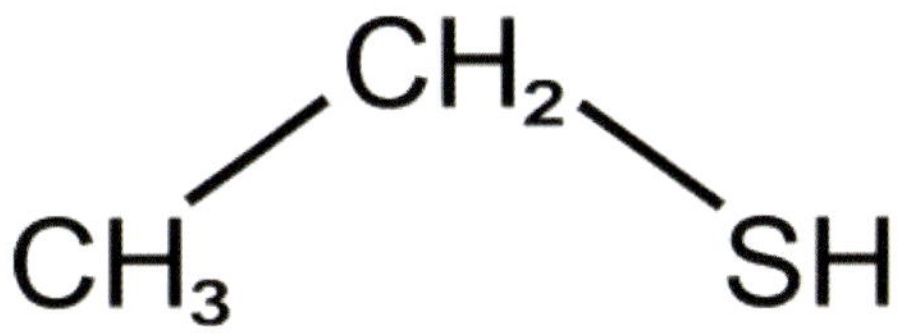

ethyl mercaptan

Mercaptans (thiols) are produced in wine by the reaction of hydrogen sulfide with other wine components such as ethanol. They can be formed if finished wine is allowed prolonged contact with the lees. This can be prevented by racking the wine. Mercaptans have a very low sensory threshold, around 1.5 μg/L, with levels above causing *onion*, *rubber*, and *skunk* type odours. Note that dimethyl disulfide is formed from the oxidation of methyl mercaptan.

Dimethyl Sulfide

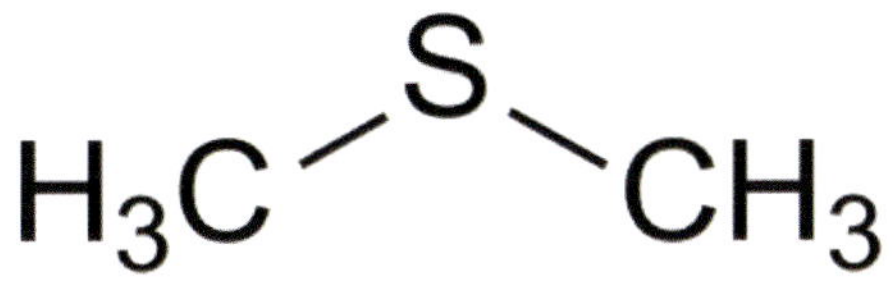

Dimethyl sulfide (DMS) is naturally present in most wines, probably from the breakdown of sulfur containing amino acids. Like ethyl acetate, levels of DMS below the sensory threshold can have a positive effect on flavour, contributing to *fruityness*, *fullness*, and *complexity*. Levels above the sensory threshold of >30 µg/L in white wines and >50 µg/L for red wines, give the wine characteristics of *cooked cabbage, canned corn, asparagus* or *truffles*.

Microbiological

Brettanomyces

The yeast *Brettanomyces* produces an array of metabolites when growing in wine, some of which are volatile phenolic compounds. Together these compounds are often referred to as *phenolic taint*, *"Brettanomyces character"*, or simply "Brett". The main constituents are listed below, with their sensory threshold and common sensory descriptors:

- 4-ethylphenol (>140 µg/L): Band-aids, barnyard, horse stable, antiseptic
- 4-ethylguaiacol (>600 µg/L): Bacon, spice, cloves, smoky
- isovaleric acid: Sweaty, cheese, rancidity

Geosmin

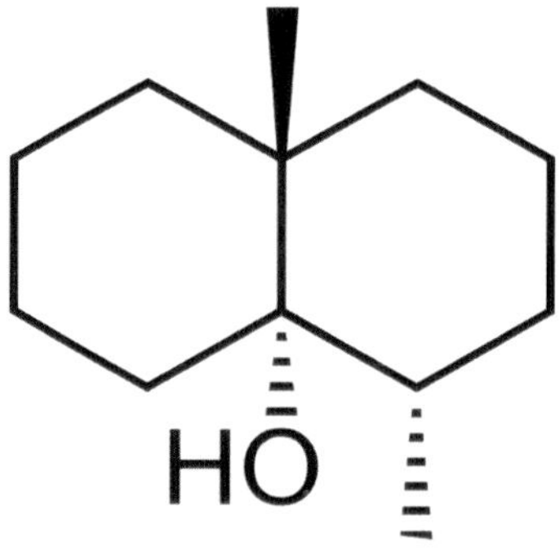

Geosmin is a compound with a very distinct *earthy, musty, beetroot*, even *turnip* flavour and aroma and has an extremely low sensory threshold of down to 10 parts per trillion. Its presence in wine is usually derived as metabolite from the growth of filamentous actinomycetes such as *Streptomyces*, and moulds such as *Botrytis cinerea* and *Penicillium expansum*, on grapes. Wines affected by but not attributed to geosmins are often thought to have earthy properties due to terroir. The geosmin fault occurs worldwide and has been found in recent vintages of red wines from Beaujolais, Bordeaux, Burgundy and the Loire in France. Geosmin is also thought to be a contributing factor in cork taint.

Lactic Acid Bacteria

Lactic acid bacteria have a useful role in winemaking converting malic acid to lactic acid in malolactic fermentation. However, after this function has completed, the bacteria may still be present within the wine, where they can metabolise other compounds and produce wine faults. Wines that have not undergone malolactic fermentation may be contaminated with lactic acid bacteria, leading to refermentation of the wine with it becoming turbid, *swampy*, and slightly effervescent or *spritzy*. This can be avoided by sterile filtering wine directly before bottling. Lactic acid bacteria can also be responsible for other wine faults such as those below.

- Bitterness Taint

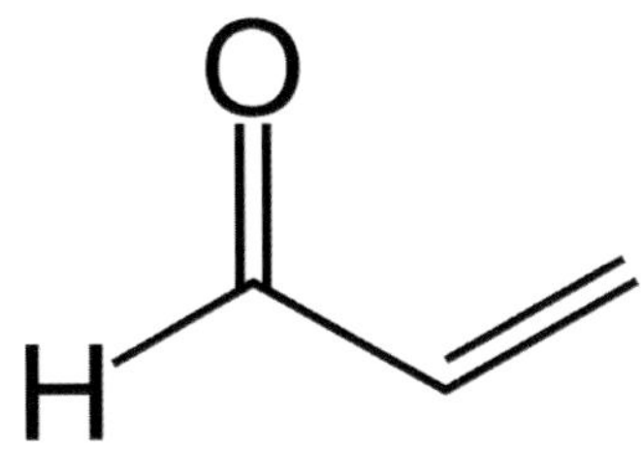

Acrolein

Bitterness taint or *amertume* is rather uncommon and is produced by certain strains of bacteria from the genera *Pediococcus*, *Lactobacillus*, and *Oenococcus*. It begins by the degradation of glycerol, a compound naturally found in wine at levels of 5-8 g/L, via a dehydratase enzyme to *3-hydroxypropionaldehyde*. During ageing this is further dehydrated to acrolein which reacts with the anthocyanins and other phenols present within the wine to form the taint. As red wines contain high levels of anthocyanins they are generally more susceptible.

- Diacetyl

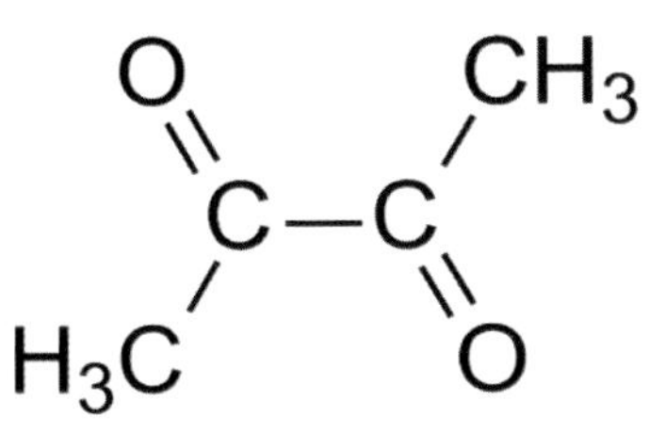

Diacetyl

Diacetyl in wine is produced by lactic acid bacteria, mainly *Oenococcus oeni*. In low levels it can impart positive *nutty* or *caramel* characters, however at levels above 5 mg/L it creates an intense *buttery* or *butterscotch* flavour, where it is perceived as a flaw. The sensory threshold for the compound can vary depending on the levels of certain wine components, such as sulfur dioxide. It can be produced as a metabolite of citric acid when all of the malic acid has been consumed. Diacetyl rarely taints wine to levels where it becomes undrinkable.

- Geranium taint

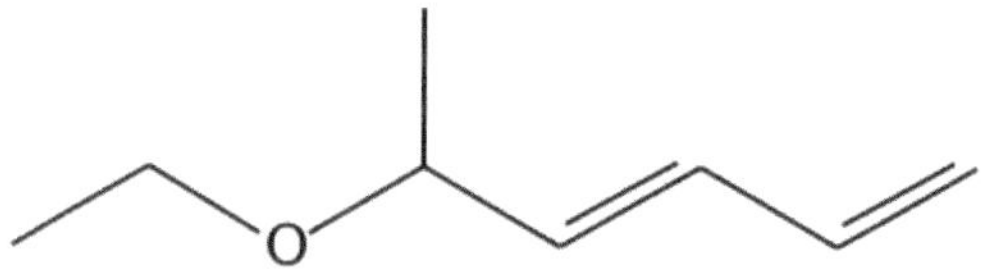

2-ethoxy-3,5-hexadiene

Geranium taint, as the name suggests, is a flavour and aroma taint in wine reminiscent of geranium leaves. The compound responsible is *2-ethoxyhexa-3,5-diene*, which has a low sensory threshold concentration of 1 ng/L. In wine it is formed during the metabolism of potassium sorbate by lactic acid bacteria. Potassium sorbate is sometimes added to wine as a preservative against yeast, however its use is generally kept to a minimum due to the possibility of the taint developing. The production of the taint begins with the conversion of sorbic acid to the alcohol *sorbinol*. The alcohol is then isomerised in the presence of acid to *3,5-hexadiene-2-ol*, which is then esterified with ethanol to form *2-ethoxy-3,5-hexadiene*. As ethanol is necessary for the conversion, the geranium taint is not usually found in must.

- Mannitol

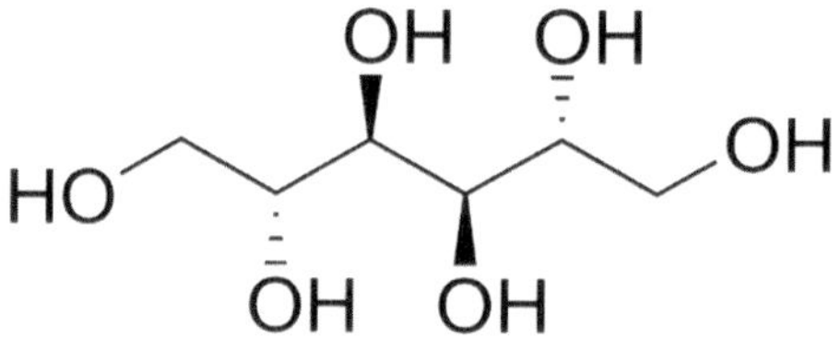

Mannitol

Mannitol is a sugar alcohol, and in wine it is produced by heterofermentative lactic acid bacteria, such as *Lactobacillus brevis*, by the reduction of fructose. Its perception is often complicated as it generally exists in wine alongside other faults, but it is usually described as viscous, ester-like combined with a sweet and irritating finish. Mannitol is usually produced in wines that undergo malolactic fermentation with a high level of residual sugars still present. Expert winemakers oftentimes add small amounts of sulfur dioxide during the crushing step to reduce early bacterial growth.

- Ropiness

Ropiness is manifested as an increase in viscosity and a *slimey* or *fatty* mouthfeel of a wine. In France the fault is known as "*graisse*", which translates to *fat*. The problem stems from the production of dextrins and polysaccharides by certain lactic acid bacteria, particularly of the genera *Leuconostoc* and *Pediococcus*.

Mousiness

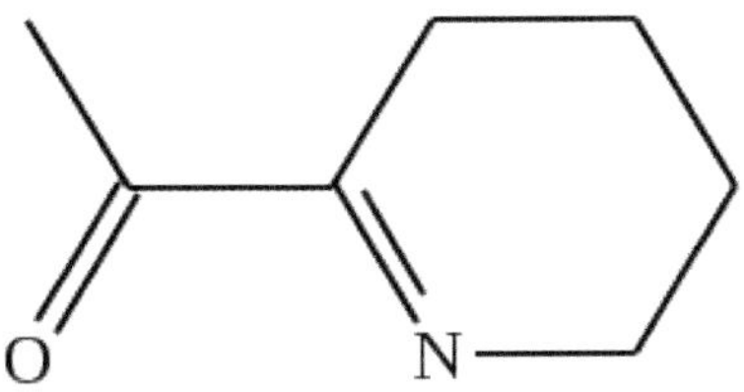

2-acetyl-3,4,5,6 tetrahydropyridine

Mousiness is a wine fault most often attributed to *Brettanomyces* but can also originate from the lactic acid bacteria *Lactobacillus brevis, Lactobacillus fermentum*, and *Lactobacillus hilgardii,* and hence can occur in malolactic fermentation. The compounds responsible are lysine derivatives, mainly;

- 2-acetyl-3,4,5,6-tetrahydropyridine
- 2-acetyl-1,4,5,6-tetrahydropyridine
- 2-ethyltetrahydropyridine
- 2-acetyl-1-pyrrolene

The taints are not volatile at the pH of wine, and therefore not obvious as an aroma. However, when mixed with the slightly basic pH of saliva they can become very apparent on the palate, especially at the back of the mouth, as *mouse cage* or *mouse urine.*

Refermentation

Refermentation, sometimes called secondary fermentation, is caused by yeasts refermenting the residual sugar present within bottled wine. It occurs when sweet wines are bottled in non-sterile conditions, allowing the presence of microorganisms. The most common yeast to referment wine is the standard wine fermentation yeast *Saccharomyces cerevisiae*, but has also been attributed to *Schizosaccharomyces pombe* and *Zygosaccharomyces bailii*. The main issues associated with the fault include turbidity (from yeast biomass production), excess ethanol production (may violate labelling laws), slight carbonation, and some coarse odours. Refermentation can be prevented by bottling wines dry (with residual sugar levels <1.0g/L), sterile filtering wine prior to bottling, or adding preservative chemicals such as dimethyl dicarbonate. The Portuguese wine style known as "vinhos verdes" relies on this secondary fermentation in bottle to impart a slight spritziness to the wine.

Bunch rots of grapes are harmful for they reduce yield and quality.

Weather conditions and control strategies influence the severity of losses which can vary between vineyards. These bunch rots can affect both wine and table grapes and control options differ depending on location, variety and trellising.

Bunch rots can be caused by several different fungi, many of which produce similar symptoms on the berries. Confusion over the cause of the bunch rot can lead to poor management decisions and crop failures.

Accurately identifying the cause of a bunch rot on your vines is critical so the correct control strategy can be implemented.

Bunch rots can affect both wine and table grapes and control options differ depending on location, variety and trellising.

References

- M. Baldy "The University Wine Course" Third Edition pgs 37-39, 69-80, 134-140 The Wine Appreciation Guild 2009 ISBN 0-932664-69-5
- DuToit, M., Pretorius, I.S. (2000). "Microbial spoilage and preservation of wine: Using weapons from nature's own arsenal - A review". South African Journal of Enology and Viticulture 21: 74-96
- Grapevine bunch rots: impacts on wine composition, quality, and potential procedures for the removal of wine faults. Steel CC, Blackman JW and Schmidtke LM, J Agric Food Chem., 5 June 2013, volume 61, issue 22, pages 5189-5206, doi:10.1021/jf400641r
- D. Bird "Understanding Wine Technology" pg 31-82, 155-184, 202-222 DBQA Publishing 2005 ISBN 1-891267-91-4

Chapter 3

Food Fermentation: Processes and Microbiology

Fermentation is an important practice in food processing. It is the process of conversion of carbohydrates into alcohol or organic acids under anaerobic conditions with the aid of microorganisms, like yeasts and bacteria. The two primary processes of lactic acid fermentation and alcoholic fermentation have been covered in extensive detail in this chapter. It also elucidates about the different kinds of fermented foods such as dairy products, sauerkraut, ngari, fermented sausages, etc.

Fermentation

Fermentation is the enzymatic decomposition and utililization of foodstuffs, particularly carbohydrates, by microbes. Fermentation takes place throughout the gastrointestinal tract of all animals, but the intensity of fermentation depends on microbe numbers, which are generally highest in the large bowel. Thus, the large intestine is quantitatively the most important site of fermention, except for species with forestomachs (ruminants). Further, there are major differences in the contribution of fermentation to energy production of different species. In carnivores like dogs and cats, and even in omnivores like humans, fermentation generates rather few calories, but in herbivores, fermentation is a way of life.

Large intestinal epithelial cells do not produce digestive enzymes, but contain huge numbers of bacteria which have the enzymes to digest and utilize many substrates. In all animals, two processes are attributed to the microbial flora of the large intestine:

- Digestion and metabolism of carbohydrates not digested in the small intestine (e.g. cellulose, residual starch)
- Synthesis of vitamin K and certain B vitamins

Cellulose is common constituent in the diet of many animals, including man, but no mammalian cell is known to produce a cellulase. Several species of bacteria in the large bowel synthesize cellulases and digest cellulose. Importantly, the major end products of microbial digestion of cellulose and other carbohydrates are volatile fatty acids, lactic acid, methane, hydrogen and carbon dioxide. Fermentation is thus the major source of intestinal gas.

Volatile or short-chain fatty acids (especially acetic, propionic and butyric acids) generated from fermentation are not only metabolized within intestinal epithelial cells, but can be absorbed by diffusion and thereby contribute fuel to systemic energy metabolism. The concentration of volatile fatty acids in the large gut is similar among mammals, but because of the enormous differences in the relative size of the large gut, the importance of microbial fermentation to energy production varies considerably among species. As examples, it has been estimated that the contribution to maintenance energy of volatile fatty acids produced in the hindgut is 6-9% in humans, 10-30% in pigs and only 2% in dogs, reflecting the relative size of their fermentation vats.

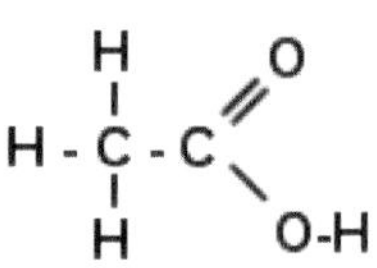

Acetic acid

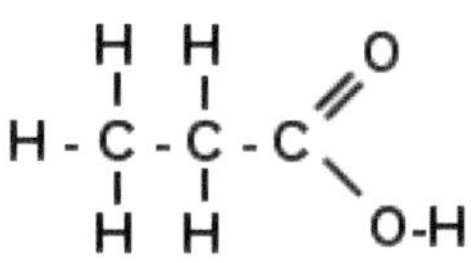

Propionic acid

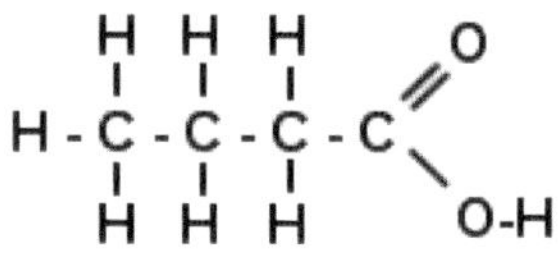

Butyric acid

Synthesis of vitamin K by colonic bacteria provides a valuable supplement to dietary sources and makes clinical vitamin K deficiency rare. Similarly, formation of B vitamins by the microbial flora in the large intestine is useful to many animals. They are not absorbed in the large intestine, but are present in feces. The behavior of coprophagy or eating feces seen particularly in rodents, rabbits and other animals is thought to be a behavioral adaption to recovery of these valuable resources.

Biological Role

Along with photosynthesis and aerobic respiration, fermentation is a way of extracting energy from molecules, but it is the only one common to all bacteria and eukaryotes. It is therefore considered the oldest metabolic pathway, suitable for an environment that did not yet have oxygen. Yeast, a form of fungus, occurs in almost any environment capable of supporting microbes, from the skins of fruits to the guts of insects and mammals and the deep ocean, and they harvest sugar-rich materials to produce ethanol and carbon dioxide.

The basic mechanism for fermentation remains present in all cells of higher organisms. Mammalian muscle carries out the fermentation that occurs during periods of intense exercise where oxygen supply becomes limited, resulting in the creation of lactic acid. In invertebrates, fermentation also produces succinate and alanine.

Fermentative bacteria play an essential role in the production of methane in habitats ranging from the rumens of cattle to sewage digesters and freshwater sediments. They produce hydrogen, carbon dioxide, formate and acetate and carboxylic acids; and then consortia of microbes convert the carbon dioxide and acetate to methane. Acetogenic bacteria oxidize the acids, obtaining more acetate and either hydrogen or formate. Finally, methanogens (which are in the domain *Archea*) convert acetate to methane.

Biochemical Overview

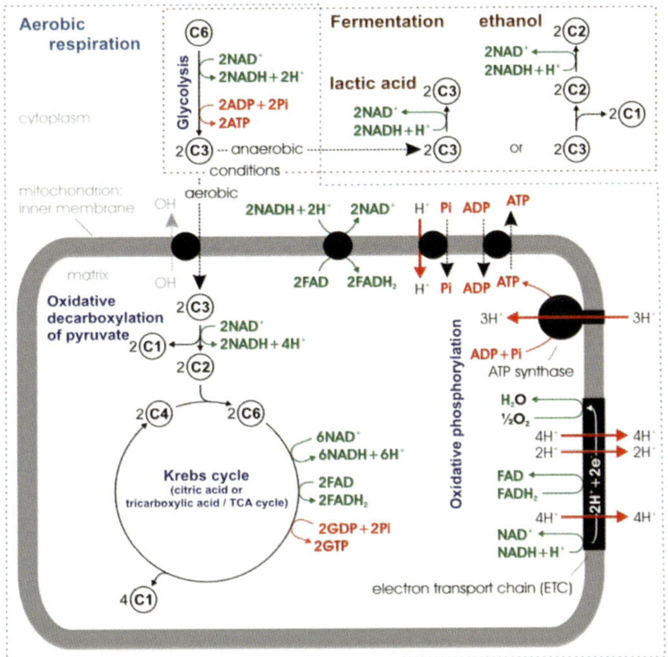

Comparison of aerobic respiration and most known fermentation types in eucaryotic cell. Numbers in circles indicate counts of carbon atoms in molecules, C6 is glucose $C_6H_{12}O_6$, C1 carbon dioxide CO_2. Mitochondrial outer membrane is omitted.

Fermentation reacts NADH with an endogenous, organic electron acceptor. Usually this is pyruvate formed from sugar through glycolysis. The reaction produces NAD+ and an organic product, typical examples being ethanol, lactic acid, carbon dioxide, and hydrogen gas (H_2). However, more exotic compounds can be produced by fermentation, such as butyric acid and acetone. Fermentation products contain chemical energy (they are not fully oxidized), but are considered waste products, since they cannot be metabolized further without the use of oxygen.

Fermentation normally occurs in an anaerobic environment. In the presence of O_2, NADH and pyruvate are used to generate ATP in respiration. This is called oxidative phosphorylation, and it generates much more ATP than glycolysis alone. For that reason, fermentation is rarely utilized when oxygen is available. However, even in the presence of abundant oxygen, some strains of yeast such as *Saccharomyces cerevisiae* prefer fermentation to aerobic respiration as long as there is an adequate supply of sugars (a phenomenon known as the Crabtree effect). Some fermentation processes involve obligate anaerobes, which cannot tolerate oxygen.

Although yeast carries out the fermentation in the production of ethanol in beers, wines, and other alcoholic drinks, this is not the only possible agent: bacteria carry out the fermentation in the production of xanthan gum.

Products

Ethanol

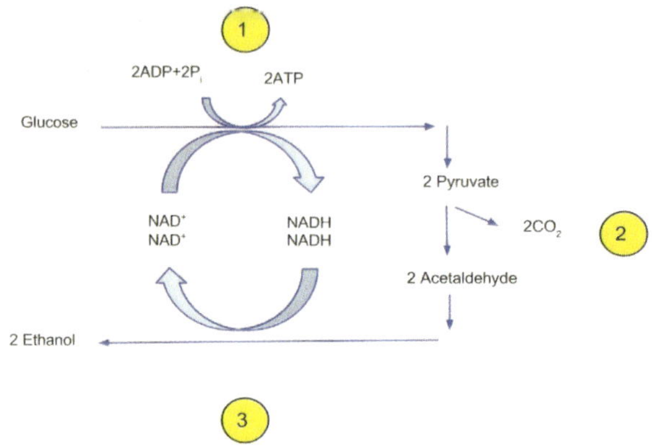

Overview of ethanol fermentation.

In ethanol fermentation, one glucose molecule is converted into two ethanol molecules and two carbon dioxide molecules. It is used to make bread dough rise: the carbon dioxide forms bubbles, expanding the dough into a foam. The ethanol is the intoxicating agent in alcoholic beverages such as wine, beer and liquor. Fermentation of feedstocks including sugarcane, corn and sugar beets produces ethanol that is added to gasoline. In some species of fish, including goldfish and carp, it provides energy when oxygen is scarce (along with lactic acid fermentation).

The figure illustrates the process. Before fermentation, a glucose molecule breaks down into two pyruvate molecules. The energy from this exothermic reaction is used to bind inorganic phosphates to ATP and convert NAD+ to NADH. The pyruvates break down into two acetaldehyde molecules and give off two carbon dioxide molecules as a waste product. The acetaldehyde is reduced into ethanol using the energy and hydrogen from NADH and the NADH is oxidized into NAD+ so that the cycle may repeat. The reaction is catalysed by the enzymes pyruvate decarboxylase and alcohol dehydrogenase.

Lactic Acid

Homolactic fermentation (producing only lactic acid) is the simplest type of fermentation. The pyruvate from glycolysis undergoes a simple redox reaction, forming lactic acid. It is unique because it is one of the only respiration processes to not produce a gas as a byproduct. Overall, one molecule of glucose (or any six-carbon sugar) is converted to two molecules of lactic acid:

$$C_6H_{12}O_6 \rightarrow 2\,CH_3CHOHCOOH$$

It occurs in the muscles of animals when they need energy faster than the blood can supply oxygen. It also occurs in some kinds of bacteria (such as lactobacilli) and some fungi. It is the type of bacteria that converts lactose into lactic acid in yogurt, giving it its sour taste. These lactic acid bacteria can carry out either homolactic fermentation, where the end-product is mostly lactic acid, or

Heterolactic fermentation, where some lactate is further metabolized and results in ethanol and carbon dioxide (via the phosphoketolase pathway), acetate, or other metabolic products, e.g.:

$$C_6H_{12}O_6 \rightarrow CH_3CHOHCOOH + C_2H_5OH + CO_2$$

If lactose is fermented (as in yogurts and cheeses), it is first converted into glucose and galactose (both six-carbon sugars with the same atomic formula):

$$C_{12}H_{22}O_{11} + H_2O \rightarrow 2C_6H_{12}O_6$$

Heterolactic fermentation is in a sense intermediate between lactic acid fermentation, and other types, e.g. alcoholic fermentation. The reasons to go further and convert lactic acid into anything else are:

- The acidity of lactic acid impedes biological processes; this can be beneficial to the fermenting organism as it drives out competitors that are unadapted to the acidity; as a result, the food will have a longer shelf life (part of the reason foods are purposely fermented in the first place); however, beyond a certain point, the acidity starts affecting the organism that produces it.
- The high concentration of lactic acid (the final product of fermentation) drives the equilibrium backwards (Le Chatelier's principle), decreasing the rate at which fermentation can occur, and slowing down growth.
- Ethanol, into which lactic acid can be easily converted, is volatile and will readily escape, allowing the reaction to proceed easily. CO_2 is also produced, but it is only weakly acidic, and even more volatile than ethanol.
- Acetic acid (another conversion product) is acidic, and not as volatile as ethanol; however, in the presence of limited oxygen, its creation from lactic acid releases additional energy. It is a lighter molecule than lactic acid, that forms fewer hydrogen bonds with its surroundings (due to having fewer groups that can form such bonds), thus is more volatile and will also allow the reaction to move forward more quickly.
- If propionic acid, butyric acid, and longer monocarboxylic acids are produced, the amount of acidity produced per glucose consumed will decrease, as with ethanol, allowing faster growth.

Hydrogen Gas

Hydrogen gas is produced in many types of fermentation (mixed acid fermentation, butyric acid fermentation, caproate fermentation, butanol fermentation, glyoxylate fermentation), as a way to regenerate NAD^+ from NADH. Electrons are transferred to ferredoxin, which in turn is oxidized by hydrogenase, producing H_2. Hydrogen gas is a substrate for methanogens and sulfate reducers, which keep the concentration of hydrogen low and favor the production of such an energy-rich compound, but hydrogen gas at a fairly high concentration can nevertheless be formed, as in flatus.

As an example of mixed acid fermentation, bacteria such as *Clostridium pasteurianum* ferment glucose producing butyrate, acetate, carbon dioxide and hydrogen gas: The reaction leading to acetate is:

$$C_6H_{12}O_6 + 4H_2O \rightarrow 2\,CH_3COO^- + 2HCO_3^- + 4H^+ + 4H_2$$

Glucose could theoretically be converted into just CO_2 and H_2, but the global reaction releases little energy.

Modes of Operation

Most industrial fermentation uses batch or fed-batch procedures, although continuous fermentation can be more economical if various challenges, particularly the difficulty of maintaining sterility, can be met.

Batch

In a batch process, all the ingredients are combined and the reactions proceed without any further input. Batch fermentation has been used for millennia to make bread and alcoholic beverages, and it is still a common method, especially when the process is not well understood.:1 However, it can be expensive because the fermentor must be sterilized using high pressure steam between batches. Strictly speaking, there is often addition of small quantities of chemicals to control the pH or suppress foaming.

Batch fermentation goes through a series of phases. There is a lag phase in which cells adjust to their environment; then a phase in which exponential growth occurs. Once many of the nutrients have been consumed, the growth slows and becomes non-exponential, but production of *secondary metabolites* (including commercially important antibiotics and enzymes) accelerates. This continues through a stationary phase after most of the nutrients have been consumed, and then the cells die.

Fed-batch

Fed-batch fermentation is a variation of batch fermentation where some of the ingredients are added during the fermentation. This allows greater control over the stages of the process. In particular, production of secondary metabolites can be increased by adding a limited quantity of nutrients during the non-exponential growth phase. Fed-batch operations are often sandwiched between batch operations.

Open

The high cost of sterilizing the fermentor between batches can be avoided using various open fermentation approaches that are able to resist contamination. One is to use a naturally evolved mixed culture. This is particularly favored in wastewater treatment, since mixed populations can adapt to a wide variety of wastes. Thermophilic bacteria can produce lactic acid at temperatures of around 50 degrees Celsius, sufficient to discourage microbial contamination; and ethanol has been produced at a temperature of 70°C. This is just below its boiling point (78°C), making it easy to extract. Halophilic bacteria can produce bioplastics in hypersaline conditions. Solid-state fermentation adds a small amount of water to a solid substrate; it is widely used in the food industry to produce flavors, enzymes and organic acids.

Continuous

In continuous fermentation, substrates are added and final products removed continuously. There are three varieties: chemostats, which hold nutrient levels constant; turbidostats, which keep cell mass constant; and plug flow reactors in which the culture medium flows steadily through a tube while the cells are recycled from the outlet to the inlet. If the process works well, there is a steady flow of feed and effluent and the costs of repeatedly setting up a batch are avoided. Also, it can prolong the exponential growth phase and avoid byproducts that inhibit the reactions by continuously removing them. However, it is difficult to maintain a steady state and avoid contamination, and the design tends to be complex. Typically the fermentor must run for over 500 hours to be more economical than batch processors.

Uses

The alcoholic beverages that can be produced by fermentation vary widely, depending primarily on two factors, the plant that is fermented and the enzymes used for fermentation. Human societies use, of course, the materials that are available to them. Thus, various peoples have used grapes, berries, corn, rice, wheat, honey, potatoes, barley, hops, cactus juice, cassava roots, and other plant materials for fermentation. The products of such reactions are various forms of beer, wine, or distilled liquors, which may be given specific names depending on the source from which they come. In Japan, for example, rice wine is known as *sake*. Wine prepared from honey is known as mead. Beer is the fermentation product of barley, hops, and/or malt sugar.

Early in human history, people used naturally occurring yeasts for fermentation. The products of such reactions depended on whatever enzymes might occur in those "wild" yeasts. Today, wine-makers are able to select from a variety of specially cultured (grown) yeasts that control the precise direction that fermentation will take.

Ethyl alcohol is not the only useful product of fermentation. The carbon dioxide generated during fermentation is also an important component of many baked goods. When the batter for bread is mixed, for example, a small amount of sugar and yeast are added. During the rising period, sugar is fermented by enzymes in the yeast, with the formation of carbon dioxide gas. The carbon dioxide gives the batter bulkiness and texture that would be lacking without the fermentation process.

Fermentation has a number of commercial applications beyond those described thus far. Many occur in the food preparation and processing industry. A variety of bacteria are used in the production of olives, cucumber pickles, and sauerkraut from raw olives, cucumbers, and cabbage, respectively. The selection of exactly the right bacteria and the right conditions (for example, acidity and salt concentration) is an art in producing food products with exactly the desired flavors. An interesting line of research in the food sciences is aimed at the production of edible food products by the fermentation of petroleum.

In some cases, antibiotics and other drugs can be prepared by fermentation if no other commercially efficient method is available. For example, the important drug cortisone can be prepared by the fermentation of a plant steroid known as diosgenin. The enzymes used in the reaction are provided by the mold *Rhizopus nigricans*.

One of the most successful commercial applications of fermentation has been the production of ethyl alcohol for use in gasohol. Gasohol is a mixture of about 90 percent gasoline and 10 percent alcohol. The alcohol needed for this product can be obtained from the fermentation of agricultural and municipal wastes. The use of gasohol provides a promising method for using renewable resources (plant material) to extend the availability of a nonrenewable resource (gasoline).

Another application of the fermentation process is in the treatment of wastewater. In the activated sludge process, aerobic bacteria (bacteria that can live without oxygen) are used to ferment organic material in wastewater. Solid wastes are converted to carbon dioxide, water, and mineral salts.

Lactic Acid Fermentation

Lactic acid fermentation is a biological process by which sugars such as glucose, fructose, and sucrose, are converted into cellular energy and the metabolic byproduct lactate.

Lactic acid fermentation is caused by some fungi and bacteria. The most important lactic acid producing bacteria is *Lactobacillus*. Other bacteria which produce lactic acid include:

- Leuconostoc mesenteroides
- Pediococcus cerevisiae

- Streptococcus lactis
- Bifidobacterium bifidus.

Lactic acid fermentation is used throughout the world to produce speciality foods:

- Western world: yogurt, sourdough breads, sauerkraut, cucumber pickles and olives
- Middle East: pickled vegetables
- Korea: kimchi (fermented mixture of Chinese cabbage, radishes, red pepper, garlic and ginger)
- Russia: kefir
- Egypt: laban rayab and laban zeer (fermented milks), kishk (fermented cereal and milk mixture)
- Nigeria: gari (fermented cassava)
- South Africa : magou (fermented maize porridge)
- Thailand : nham (fermented fresh pork)
- Philippines : balao balao (fermented rice and shrimp mixture).

The presence of lactic acid, produced during the lactic acid fermentation is responsible for the sour taste and for the improved microbiological stability and safety of the food. This lactic acid fermentation is responsible for the sour taste of dairy products such as cheese, yoghurt and kefir.

Lactic acid fermentation also gives the sour taste to fermented vegetables such as traditionally cultured sauerkraut and pickles. The sugars in the cabbage are converted into lactic acid and serve as a preservative.

Yogurt Fermentation

Yogurt is made by fermenting milk with friendly bacteria, mainly Lactobacillus bulgaricus and Streptococcus thermophilus. Yogurt fermentation was invented probably by accident by Balkan tribes thousands of years ago. Yogurt remained mainly a food of eastern Europe until the 1900s, when the biologist Mechnikov created the theory that lactobacillus bacteria in yogurt are responsible for the unusually long lifespans of the Bulgar people. The milk sugar or lactose is fermented by these bacteria to lactic acid which causes the characteristic curd to form. The acid also restricts the growth of food poisoning bacteria. During the yogurt fermentation some flavours are produced, which give yogurt its characteristic flavour. Yoghurt can easily be made at home using a live yogurt as the starter culture. To make you own yogurt use the following process. Bring the milk (or soymilk) to boiling point and cool down to 40-45°C. Pour this milk in a sterile container and and per liter milk about 100 ml live yogurt. Mix with a sterile spoon and incubate at 40-44°C during 4 to 6 hours or until the yogurt is set. Put the yogurt in the refrigerator. If you worked under hygienic conditions, you can use your own yogurt as a starter for your next batch.

Magou Fermentation

Magou is very popular in South Africa, especially among the Bantu people. Magou is a lactic acid fermented porridge made from maize. To make magua a 10 percent maize meal slurry is cooked, cooled and inoculated with wheat flour, which contains the bacteria. Magou is also produced on industrial scale and is then packed in cartons. In the industrial process the magou is inoculated with *lactobacillus delbreuckii* cultures.

Kefir Fermentation

Kefir fermentation is similar to yogurt fermentation. Yogurt is only fermented by bacteria but kefir fermentation involves the help of bacteria as well as yeasts. These yeast produce some alcohol and carbon dioxide, which gives kefir its typical fizzy aspect. Kefir is inoculated with special kefir grains. These grains are mixtures of bacteria and yeasts in a matrix of proteins, lipids and carbohydrates. Kefir fermentation is done at room temperature, which makes the process easier. On the other hand, not everyone likes the taste of kefir.

Alcoholic Fermentation

Alcohol fermentation is a type of anaerobic respiration. Fermentation can be defined as a process that releases energy from sugars or other organic molecules. It does not require oxygen (but sometimes can occur in its presence). It does not require the use of the Krebs cycle or an electron transport chain. It uses an organic molecule as the final electron acceptor. Alcohol fermentation begins with the glycolysis of a molecule of glucose to yield two molecules of pyruvic acid. The two molecules of pyruvic acid are then reduced to two molecules of ethanol. The ethanol and CO_2 are the final products of alcohol fermentation.

Alcohol Fermentation is Carried out by a Number of Bacteria and Yeasts.

Alcohol fermentation is also a biotechnological process. In this process we use yeasts *Saccharomyces* as bioculture and water solution of monosaccharides as the culture media.

The quantity and the sort of monosaccharides influence the efficiency of alcohol fermentation.

Yeasts involved in the production of alcoholic beverages are capable of facultative anaerobic growth. If given access to oxygen, yeasts perform aerobic respiration to metabolize carbohydrates to CO_2 and water. Denied oxygen they ferment carbohydrates and produce ethanol and CO_2.

Alcoholic fermentation - or primary fermentation - is a fundamental process for the production of wine. Its role is not only limited to the transformation of sugar into alcohol and carbon dioxide - as well as other byproducts - indeed, it is what defines, or better to say, reveals, the organoleptic qualities of the grape. In fact, all the potential qualities of a wine are, like to say, hidden in the grape: they will be revealed or they will disappear forever by means of alcoholic fermentation and during the other phases of vinification. The role of alcoholic fermentation is fundamental for the development of aromatic and gustatory qualities of a wine, it is not by chance it is said "primary aromas are hidden in the grape, but it is fermentation which will reveal them".

Steel tanks for fermentation

Fermentation is a natural process allowing the transformation of grape juice - the must - into wine. Alcoholic fermentation, or primary fermentation, can be considered as a chain reaction of chemical phenomena. During these processes, the sugar contained in the grape juice is converted by yeast's enzymes into ethyl alcohol and carbon dioxide. According to the type of yeast, from the conversion of sugar is obtained about 50% of alcohol, 45% of carbon dioxide, 3% of glycerol and 2% of other byproducts. This complex phenomenon, unknown for centuries, produces other substances as well, considered as "byproducts" and which play an essential role in the aromatic and gustatory qualities of wine. Of these products, the most important ones are acetaldehyde, acetic acid, ethyl acetate, glycerol and other types of alcohol. These "secondary" products are in fact essential for the development of any wine's aromatic qualities and, depending on how the alcoholic fermentation was done, aromas can get refined or spoiled.

Uses of Alcohol Fermentation

Fermentation has been around for centuries. It has been used in many ways to perform various tasks. Alcohol fermentation in particular has carried out the most useful tasks. Without alcohol fermentation we would not have the bread-making, brewing, or wine-making industries as we do today.

The Alcoholic Beverage Industries

The alcoholic beverages that can be produced by fermentation can vary depending on two major factors: the plant that is being fermented and the enzymes used for fermentation. Previously plants such as grapes, berries, corn, rice, wheat, honey, potatoes, barley, hops, cactus juice, casava roots, and other plants have been used during fermentation. These plants can result in different types of beer, wine, or distilled liquors. Before cultured yeasts were made available, people used naturally occuring yeasts to conduct fermentation, but as times have changed industries can now choose from a wide variety of yeasts to control the precise direction in which fermentation will go.

Brewing

In the process of the brewing barley, hops, and malt sugar are fermented to create beer. As you read above industries can now choose certain yeasts to carry out fermentation. Using this technology the brewing companies are able to make a unique taste depending on the yeast they choose. Different yeasts can also determine the percentage of alcohol that is able to build up in the large vats, where the fermentation takes place, before fermentation comes to an end (usually 3-6%). Beers are usually bottled after fermentation has ceased, and some carbon dioxide is added to the bottle right before it is capped.

Wine-Making

There are several types of wines that can be created with different types of plants and yeasts. Some wines can be made with rice, like sake in Japan, or honey, like mead, but most wines are made with fermented grapes. In the wine-making industry several different yeasts can be chosen to control the the direction of fermentation. These yeasts can also determine the percentage of alcohol which can build up in the vats before fermentation ceases (usually 10-20%). Most wines are bottled after fermentation has ended, but for some wines like champagne it is bottled before and some carbon dioxide is then bottled with the drink.

Bread-Making

Ethyl alcohol is not the only useful product of fermentation, carbon dioxide is very important in the bread-making industry. When the batter for bread is made a small amount of sugar and yeast is added to the mix. During the rising period, sugar is fermented by enzymes in the yeast with the formation of carbon dioxide gas. The carbon dioxide gives the batter texture and bulkiness that would be lacking without the process of fermentation.

Other Uses

Alcohol fermentation is also used in the production of olives, cucumber pickles, sauerkraut, some antibiotics, and the production of gasohol.

Fermented Foods

Fermented foods are foods that have been through a process of lactofermentation in which natural bacteria feed on the sugar and starch in the food creating lactic acid. This process preserves the food, and creates beneficial enzymes, b-vitamins, Omega-3 fatty acids, and various strains of probiotics.

Natural fermentation of foods has also been shown to preserve nutrients in food and break the food down to a more digestible form. This, along with the bevy of probiotics created during the fermentation process, could explain the link between consumption of fermented foods and improved digestion.

Cultures around the world have been eating fermented foods for years, from Sauerkraut in Germany to Kimichi in Korea and everywhere in between. Studies have even shown the link between probiotic rich foods and overall health.

Fermented foods are "the flavourful space between fresh and rotten".

The process of fermenting foods is not new one, evidence indicates that early civilizations were making wine and beer between 7,000 and 8,000 years ago, and bread even before that. India is traditionally rich in fermented foods. In the Indian sub-continent, fermented food using local food crops and other biological resources are very common. Fermented foods such as idli and dahi were described as early as 700 BC. At present, there are hundreds of fermented foods with different base materials and preparation methodology. Each fermented food is associated with a unique group of microbiota, which increases the level of proteins, vitamins, essential amino acids and fatty acids.

Benefits: Foods like idlis, dosas, dhoklas, wadas and kadhi are some of the lactobacillus fermented cereals and legumes that are commonly consumed in India. The fermented foods increase the absorption of vital minerals from gastrointestinal tract, thus preventing mineral deficiencies. Bread, fish sauce, wine and beer are some of the yeast-based fermented food beverages. Experts say that probiotic foods improve immunity, aids in better digestion, absorption of calcium and prevention of allergies. They also help in treating diarrhoea. If a pregnant woman consumes probiotic food like a month before her delivery, this will help prevent the new born infant from contracting allergies to some extent. Also, if infants who are six months and older are given probiotic food it helps to build their immunity.

Indian fermented foods are consumed by the local population not just as a diet but as traditional medicine too. Many of the foods were observed to have beneficial effect during ailment by the local people and they are used as a special diet or medicine for ages. Fermented food idli is easily digested and often used as food for infant and invalids. Fermented milk dahi can be used to cure intestinal disease such as diarrhea; intake of dahi has anti-cholesteremic, anticarcinogenic, anti-diabetic, angiotensin-converting enzyme inhibition effect and anti-atopic dermatitis effect. The above properties are validated by in vivo experiments. Certain fermented vegetable products such as sinki and common pickles are said to be good appetizers, and the ethnic people use these foods for remedies from indigestion. Bhutia women use gundruk soup and give it to breast-feeding mothers to improve milk efficiency. It is considered as a tonic for old age people. Fermented radish root pieces are called sinki. It is very effective in curing diarrhea and stomach pain, and is consumed mostly during lean period. Fermented rai helps in curing stomach pain and gas trouble, and significantly improves digestion. Kanjika or kanji is a lactic fermented rice product. It has been prescribed for a number of chronic diseases by Indian Ayurvedic practitioners.

Carrot Kanji is considered to have high nutritional value and cooling and soothing properties. Beetroot kanji is considered to have potential to prevent infection and malignant disease. Fermented BS juice has preservation property similar to vinegar. The Meitei community in Northeastern India use fermented BSs (Soibum) along with fermented fish against the plague disease. Ngari-dry fish of Puntus breed Phutunis or Eromus dandricus is mixed with Iromba (fermented BSs) and this dish called Yongchak Iromba is considered as an appetizer and is well known to the old age women of the Meitei community. Most of the claims are based on traditional beliefs; hence there is a scope to conduct experiment-based research to prove these claims.

Examples of Some Indian Fermented Foods.

Fermented food	Ingredients	Place of origin	Related Microorganisms
Rabdi (rabadi)	Flour of barley, pearl millet, corn or soybean and country buttermilk	Rajasthan	*Bacillus* and *Micrococcus* sp.
Kulu	Wheat flour, buttermilk	Himachal Pradesh	*Lactobacillus* sp.
Idli	Rice, black gram dhal, table salt, fenugreek seeds	South India	*L. mesenteroides, E. faecalis, P. cerevisiae*
Dosa	Rice, black gram dhal (either raw or parboiled rice), table salt	South India	*L. mesenteroides, E. faecalis*
Dhokla	Bengal gram dhal, rice and leafy vegetables	Gujrat	*L. fermentum, L. mesenteroides, E. faecalis*

Chilra or lwar	Wheat/barley, buckwheat flour and starter material Treh	North India	Not reported
Sinki	Radish root	North-east India	*L. casei, L. brevis, L. plantarum, L. fallax, L. fermentum*
Kinema	Soybeans	Darjeeling, Sikkim	*E. faecium*
Kanji	Carrot or beet root, rice, mustard	North India	*L. pentosus, L. paraplantarum, L. plantarum*
Curd (Dahi)	Milk	India	*S. cremoris, S. lactis, S.thermophilus, L. bulgaricus, L. acidophilus, L. helveticus, L. cremoris,, Lactobacillus delbrueckii subsp. indicus*
Gundruk	Leaves of mustard/radish/cauliflower	Arunachal Pradesh	*P. pentasaceous, L. fermentum, L. casei*

Food Poisoning: Except above merits, fermented foods are more prone to be contaminated. Over growth of unwanted secondary microbial species may alter the fermented food properties by releasing several harmful metabolites and lead to the intoxication or poisoning. Numerous food-borne outbreaks have been reported throughout the India. Due to the contamination, a popular nutritive Indonesian fermented food "*Tempeh*" has been periodically caused food poisoning and death in Central Java. Microbes such as *Clostridium, Brucella, Escherichia, Salmonella* and *Bacillus* species are generally associated with food borne illness. Therefore, special caution should be taken before consuming the fermented food products.

Ngari

Among the many fermented foods, such as hawaijar, soibum, soidon, hentak, ziang sang and atingba, ngari is a popular fermented fish product.

Ngari is indigenous to Manipur and is also valued as a food ingredient in other states of north-east India. Fermentation is one of the oldest and most economical methods for Ngari -a traditional fish product of Manipur, India Fermented foods and beverages are popular among the people of the north-eastern states of India. Ngari, a fermented fish product made from sun-dried non-salted fish, is one such food item well liked by the people of Manipur. producing and preserving foods. In addition to preservation, fermented foods can also have the added benefits of enhancing flavour, increasing digestibility and improving nutritional and pharmaceutical value. The popularity of the product is due to its inviting flavour. The people of Manipur consume it almost every day; hardly a day passes without ngari being cooked. It is one of the most important ingredients for an oil-free food preparation known as 'eromba' and 'kangsoi' in Manipuri cuisine. It has several local names in the north-east, such as seedal, sepaa, hidal and shidal in Assam,

Tripura, Mizoram, Arunachal Pradesh and Nagaland while in Manipur, it is called ngari. For preparation of ngari, a sun-dried non-salted fish, Phoubu nga (Puntius sophore) is fermented by a traditional method in the valley region of Imphal, the capital of Manipur state, and in the surrounding areas. The colour of the best quality product is dull white which gradually becomes light to deep brownish on continuous exposure to air.

Fermented Sausages

Fermented sausages are a class of chopped or ground meat products that, as a result of microbial fermentation of a sugar have reached a pH of 5.3 or lower and have undergone a drying/aging process to remove 15-25% moisture. They are classified according to moisture content—dry (e.g., Pepperoni, Salami) or semi-dry (e.g., Summer sausage, Thuringer, Cervelat, Landjaegar).

After mixing the ingredients and stuffing into casings, sausages are placed in an environment with controlled temperature and humidity. At this fermentation stage, the pH of the sausages decreases due to lactic acid bacteria that produce lactic acid from metabolizing sugar. Lactic acid bacteria such as Lactobacillus or Pediococcus are either naturally present or added as starter cultures. Aside from bacteria, chemical acidulants such as glucono-delta-lactone and citric acid can also be added to rapidly increase acidity and create an extra margin for safety. The pH drop causes the proteins to give up water, resulting in a drying effect that creates an environment unfavorable to spoilage organisms. Drying continues after the fermentation stage and more moisture is removed from the sausage.

Fermented sausages should attain a pH of 5.3 or lower within the proper time frame in order to control the growth of pathogenic bacteria such as E. coli 0157:H and Staphylococcus aureus. During the fermentation of sausages to a pH 5.3, it is necessary to limit the time during which the sausage is exposed to temperatures exceeding 15.6°C, otherwise the product will spoil, even though the recommended pH was attained. This time frame is temperature dependent and these are the criteria (tables shown are guidelines in developing fermentation process):

Time in C degree-hours above 15.6°C	Maximum Chamber Temperature
Less than 665	Less than 33°C
<555	32-37°C
<500	Greater than 37°C
Degrees are measured as the excess over 15.6° C, the critical temperature at which staphylococcal growth effectively begins.	

The pH of every batch of sausages should be periodically checked and recorded during fermentation to ensure that the pH reaches 5.3 or lower within the required period. The

LAQUAtwin pH meter can be used by home sausage-makers or meat processing plants for monitoring the pH of their sausages.

Constant Temperature Fermentation		
Degree (C)-hours limit for the corresponding temperature	Chamber Temperature °C	Maximum hours to pH 5.3
665	24	78.9
665	26	63.8
665	28	53.6
665	30	46.2
665	32	40.5
555	33	31.8
555	34	30.1
555	35	28.6

This table provides maximum hours that a product may be fermented at a given constant fermentation temperature to obtain pH 5.3. Example: At 26°C constant temperature, a sausage must reach pH 5.3 within 60 hours or less. Those hours can also be calculated for any temperature.

Method

Test the meter's accuracy regularly and if necessary, calibrate the LAQUAtwin pH meter using 4.01 and 7.00 pH buffers according to manufacturer's instructions.

Sample Measurement

This method is based on the Australia New Zealand Food Standards Code - Standard 1.6.2.

1. Mince a representative portion of the sausage and place that portion in a stoppered bottle with twice its weight of water then shake.

2. Place a portion of the liquid into the sensor. Record the pH and temperature once stabilized.

3. Clean the sensor with soap and warm water to remove oily residues and/or protein cleaning solution to remove proteins. Rinse the sensor with water and blot dry with soft tissue.

To obtain accurate results, a uniform temperature should be maintained for the standard buffer solutions and samples. The test should be made at a temperature between 20-30°C, the optimum is 25°C.

Average Characteristics of Australian UCFM processes and products*		
Characteristic	Salami (range of variation)	Mettwurst (range of variation)
Composition (lean % : fat %)	80.4 : 19.6 (70:30 - 90:10)	83.25 : 16.75 (70:30 - 96:4)
NaCl (%)	2.45 (2.0 - 3.3)	2.02 (1.3 - 2.8)
Nitrite (ppm)	284 (145- 490)	211 (35 - 490)
Final pH	4.72 (5.0 - 4.4)	4.66 (4.8 - 4.4)
Fermentation time (hrs)	49 (24 - 72)	42 (18 - 72)
Fermentation temp. (°C)	23.3 (18 - 28)	28.6 (17 - 40)
Ripening time (days)	14.8 (1-30)	5.3 (0 – 28)
Ripening temp. (°C)	14.1 (4 - 32)	18 (0 - 40)

Results and Benefits

Fermented sausages such as salamis have been associated with food-poisoning outbreaks globally due to pathogenic microorganisms. In light of this, it is of utmost importance to be in strict compliance with the regulatory guidelines for fermented sausages. In the USA, the Food Safety and Inspection Service requires that the shelf-stable dry sausages be nitrite cured, fermented, smoked, reached a final pH of 5 or less and have a moisture/protein ratio of 1.9:1 or less. The two main factors contributing to the safety and stability of these products are low pH and reduced water activity. The product must reach pH 5.3 or below (or other validated pH) within a specified period to control the growth of pathogenic microorganisms including Staphylococcus aureus and pathogenic E. coli.

Idli/Dosa

Idli or idly (pronunciation (help·info)) (/ɪdliː/) are a type of savoury rice cake, originating from the Indian subcontinent, popular as breakfast foods throughout India and northern Sri Lanka. The cakes are made by steaming a batter consisting of fermented black lentils (de-husked) and rice. The fermentation process breaks down the starches so that they are more readily metabolized by the body.

Idli have several variations, including rava idli, which are made from semolina. Regional variants include *sanna* of Konkan and Enduri Pitha of Odisha.

Preparation

To make Idli, four parts uncooked rice (Idli rice or parboiled rice) to one part whole white lentil (urad dal, Vigna mungo) are soaked separately for at least four hours to six hours or overnight. Optionally spices such as fenugreek seeds can be added at the time of soaking for additional flavour. Once done soaking, the lentils are ground to a fine paste and the rice is separately coarsely ground, then they are combined. Next, the mixture is left to ferment overnight during which its volume will more than double. After fermentation some of the batter may be kept as a starter culture for the next batch. The finished idli batter is put into greased moulds of an idli tray or "tree" for steaming. The perforated molds allow the idlis to be cooked evenly. The tree holds the trays above the level of boiling water in a pot, and the pot is covered until the idlis are done (about 10–25 minutes, depending on size). A more traditional method is to use leaves instead of moulds.

Serving

Since plain idlis are mild in taste, a condiment is considered essential. Idlis are often served with sambar but this varies greatly by region and personal taste. Idlis are also frequently served with chutnies (coconut based), kaara chutney (onion based) or Spicy Fish Curries. The dry spice mixture podi is convenient while travelling.

Idli and medu vada served with sambar, green chutney, and red chutneys on banana leaf

Idli and vada served with separate sambar and chutney

Idlis with ghee, ginger pickle and coconut chutney

Variations

With the emigration of south Indians and Sri Lankans throughout the region and world, many variations on idli have been created in addition to the almost countless local variations. Hard-to-get ingredients and differing cooking customs have required changes in both ingredients and methods. Parboiled rice can reduce the soaking time considerably. Store-bought ground rice or cream of rice may also be used. Similarly, semolina or cream of wheat may be used for preparing rava idli (wheat idli). Dahi (yogurt) may be added to provide the sour flavor for unfermented batters. Prepackaged mixes allow for almost instant idlis.

In addition to or instead of fenugreek other spices may be used such as mustard seeds, chili peppers, cumin, coriander, ginger, etc. Sugar may be added to make them sweet instead of savory. Idli may also be stuffed with a filling of potato, beans, carrot and masala. Leftover idlis can be cut-up or crushed and sautéed for a dish called idli upma. You can even prepare dahi idli by combining curd and idli together and then tempering the dish. Rather than a stovetop steamer, microwave and automatic electric idli steamers are available with convenient non-stick coating. Batter preparation using a manual rocking rock grinder can be replaced by electric grinders or blenders (mixies). Many restaurants have also come up with fusion recipes of Idlis as idly manchurian, idly fry, chilly idly and a lot of different ideas.

Muday idli is a Mangalorean variant of idli.

Rave idli (Sooji idli) is a specialty of Karnataka

Ragi idli - Ragi flour mixed with small portion of Rice flour, Urad dal.

Dosa

Dosa is a type of pancake from the Indian subcontinent, made from a fermented batter. It is somewhat similar to a crepe in appearance. Its main ingredients are rice and black gram. Dosa is a typical part of the Southern Indian diet and popular all over the Indian subcontinent. Traditionally, dosa is served hot along with sambar, a stuffing of potatoes, and chutney. It can be consumed with idli podi as well.

Dosa with Sambar and various Chutnies in restaurant

This is one of the many varieties of Dosas prepared in India and served for breakfast. It's made with ground tomatoes and seasoning

Nutrition

Dosa with chutney and sambar traditionally served in banana leaf

Dosa is high in carbohydrates and contains no added sugars or saturated fats. As its key ingredients are rice and black gram, it is also a good source of protein. The fermentation process increases the vitamin B and vitamin C content. There are also instant mix products for making dosa which usually contain higher amounts of rice.

Preparation

A mixture of rice and black grams that has been soaked in water is ground finely to form a batter. Some add a handful of fenugreek seeds soaked along with the rice. The proportion of rice to lentils is generally 4:1 or 5:1. The batter is allowed to ferment overnight. After the overnight fermentation, the batter is mixed with water to get the desired thickness. The batter is then ladled onto a hot *tava* (griddle) greased with oil or *ghee* (clarified butter). It is spread out evenly with the base of a ladle or bowl to form a pancake. A dosa is served hot, either folded in half or rolled like a wrap. It is also usually served with chutney and sambar. The mixture of black grams and rice can be replaced with highly refined wheat flour or semolina.

Batter being spread uniformly

After being cooked for some time

Serving

Dosa can be stuffed with fillings of vegetables and sauces to make a quick meal. They are typically served with a vegetarian side dish which varies according to regional and personal preferences. Common side items are:

- Sambar
- Chutney: examples include coconut chutney (a semi-solid paste made up of coconut, dal (lentils), green chilli and mint or coriander)
- There are varieties of chutney served along with Dosa
- Idli podi or molagapodi: a powder of spices and sometimes desiccated coconut mixed with sesame oil or ghee
- Indian pickles

Variations

The most popular version is the Masala Dosa, with a filling of the potato masala. Mysore Masala is the spicier version of it. Sada (plain) is without filling, paper dosa is a thin and crisp version. Rava dosa is made crispier using semolina. Newer recipes have been developed that use fusion.

Though dosa typically refers to the version made with rice and lentils, many other versions exist.

Types of Dosa	
Name	**Key ingredients**
Masala dosa	spiced potatoes tucked inside the dosa with red chutney smeared over the dosa.
Plain Dosa	Dosa served with only chutney and sambar and no filling.
Paneer dosa	spiced paneer filling inside the dosa.
Palak dosa	layered with palak (spinach) paste inside the folds of dosa.
Mini soya dosa	soya milk and wheat flour
Pesarattu (Green Dosa)	green gram
Light white dosa	rice and coconut
Mysore Masala Dose	rice, black gram, fenugreek seeds
Onion Rava Dosa	Semolina, rice flour
Ragi Wheat Dosa	Ragi, Whole Wheat Flour
Rava dosa	rava or sooji
Benne Dose	butter
Neer dosa	watery rice batter
Amboli, Ghavan, Dhirde	In coastal parts of Maharashtra, variations known as amboli, ghavan and dhirde (or Dhirade) exist. Amboli and Ghavan (like dosa) are thin rice crêpes prepared with fermented batter, while dhirde is prepared with unfermented batter.

Paper Roast, a wafer thin crispy dosa, served in restaurants.

Ghee Roast

Wheat Flour Dosa

Masala Dosa

Masala dosa as served in Kuala Lumpur, Malaysia

The *masala dosa* is made by stuffing a dosa with a lightly cooked filling of potatoes, fried onions and spices. The dosa is wrapped around an onion and potato curry or masala. Masala dosa was listed as number 49 on *World's 50 most delicious foods* compiled by *CNN Go* in 2011.

Related Foods

- Uttapam: a dosa-like dish made from the same batter. Unlike a dosa, which is crisp and unlike relatively soft crepes, it is a thick pancake mostly topped with diced onions, tomatoes, cilantro or cheese. Uttapam is sometimes characterized as an Indian pizza.
- Pesarattu: In Andhra Pradesh a dosa-like preparation prepared from green gram, which is typically served with a ginger and tamarind chutney
- Adai: a dosa-like dish prepared from a combination of Toor Dal, rice, curry leaves, red chillies and asafoetida. The batter is not fermented. Usually eaten with Jaggery or Aviyal.
- Appam/aappam/hopper : a pancake prepared from a combination of patted rice batter. The center is thicker and the outer rim is very thin. Served with sweet coconut milk.
- Chakuli pitha: batter contains more black gram and less rice flour
- Apam balik: made from a mixture of flour, eggs, sugar, baking soda, coconut milk and water.
- Jianbing: dosa like dish made in china

Sauerkraut

Sauerkraut is pickled or fermented cabbage that has a distinctive sour flavor.

Sauerkraut is a type of fermented cabbage with major health benefits.

It's thought to have originated in China more than 2,000 years ago. Back then, fermentation was one of the methods used to keep foods from spoiling quickly.

Sauerkraut survived the test of time to become a popular side dish and condiment in many cultures. It's especially appreciated in Germany, where its name comes from.

Due to the fermentation it undergoes, sauerkraut offers nutrition and health benefits far beyond those of fresh cabbage.

Sauerkraut is Very Nutritious

Sauerkraut contains many nutrients important for optimal health. One cup (142 grams) provides you with:

- Calories: 27
- Fat: 0 grams
- Carbs: 7 grams
- Fiber: 4 grams
- Protein: 1 gram
- Sodium: 39% of the RDI
- Vitamin C: 35% of the RDI
- Vitamin K: 23% of the RDI
- Iron: 12% of the RDI
- Manganese: 11% of the RDI
- Vitamin B6: 9% of the RDI
- Folate: 9% of the RDI
- Copper: 7% of the RDI
- Potassium: 7% of the RDI

Sauerkraut is particularly nutritious because it is fermented.

Sauerkraut fermentation is the process of microorganisms on the cabbage digesting its natural sugars and converting them into carbon dioxide and organic acids.

Fermentation starts when yeast and bacteria that are naturally present on the cabbage, your hands and in the air come into contact with the sugars in the cabbage.

Sauerkraut fermentation creates conditions that promote the growth of beneficial probiotics, which are also found in products like yogurt and kefir.

Probiotics are bacteria that provide powerful health benefits. They also help make foods more digestible, which increases your gut's ability to absorb the vitamins and minerals they contain. This is what makes sauerkraut more nutritious than raw cabbage or coleslaw.

However, unlike cabbage, sauerkraut can be high in sodium. Keep this in mind if you're watching your salt intake.

Improves your Digestion

Your gut is said to contain over 100 trillion microorganisms or "gut flora," which is more than 10 times the total number of cells in your body.

Sauerkraut contains probiotics, which are beneficial bacteria that act as the first line of defense against toxins and harmful bacteria. They can also improve your digestion and overall health.

Probiotics like those in sauerkraut can help improve the bacterial balance in your gut after it's been disturbed by the use of antibiotics. This can help reduce or prevent antibiotic-provoked diarrhea.

Research also shows that probiotics help reduce gas, bloating, constipation, diarrhea and symptoms linked to Crohn's disease and ulcerative colitis.

Most probiotic-rich foods and supplements contain 5–10 billion colony-forming units (CFUs) per serving. In comparison, 1 cup of sauerkraut contains around 3 billion CFUs.

However, according to some non-scientific sources, some types of sauerkraut may contain much higher amounts.

Different probiotic strains may also provide different advantages. Thus, consuming a wide variety of strains may give you a broader range of health benefits.

On this aspect, sauerkraut may have the advantage. Research has reported that one serving may contain up to 28 distinct bacterial strains.

Like most other fermented foods, sauerkraut also contains a variety of enzymes, which help break down nutrients into smaller, more easily digestible molecules.

Boosts your Immune System

Sauerkraut is a rich source of immune-boosting probiotics and nutrients.

For starters, the bacteria that populate your gut can have a strong influence on your immune system. The probiotics found in sauerkraut help improve the balance of bacteria in your gut, which keeps your gut lining healthy.

A stronger gut lining helps prevent unwanted substances from "leaking" into your body and causing an immune response.

Maintaining a healthy gut flora also helps prevent the growth of harmful bacteria and may even boost the production of natural antibodies.

Moreover, regularly consuming probiotic-rich foods like sauerkraut may reduce your risk of developing infections such as the common cold and urinary tract infections.

If you do get sick, regularly consuming probiotic-rich foods may help you recover faster and could reduce your chance of needing antibiotics by around 33%.

In addition to being a rich source of probiotics, sauerkraut is also rich in vitamin C and iron, both of which contribute to a healthy immune system.

In particular, upping your vitamin C intake when you have the common cold may help you get rid of symptoms more quickly.

May Help you Lose Weight

Regularly consuming sauerkraut may help you lose weight and keep it off.

That's partly because sauerkraut, like most vegetables, is low in calories and high in fiber. High-fiber diets keep you fuller for longer, which may help you naturally reduce the number of calories you eat each day.

Sauerkraut's high probiotic content may also contribute to a trimmer waistline.

The exact reasons aren't yet fully understood, but scientists believe that probiotics may have the ability to reduce the amount of fat your body absorbs from your diet.

Various studies report that participants given probiotic-rich foods or supplements lost more weight than those given a placebo.

A recent study even reports that purposely overfed participants given probiotics gained about 50% less body fat than overfed participants given a placebo. This suggests that a probiotic-rich diet may even help prevent weight gain.

However, these results are not universal. In addition, different probiotic strains may have different effects. Thus, more research is needed to determine the effectiveness of sauerkraut-specific probiotic strains on weight loss.

Helps Reduce Stress and Maintain Brain Health

While your mood can affect what you eat, the reverse is also thought to be true. What you eat can affect your mood and brain function.

An increasing number of studies are discovering an intimate connection between your gut and brain.

They've found that the type of bacteria present in your gut may have the ability to send messages to your brain, influencing the way it functions and perceives the world.

For instance, fermented, probiotic-rich foods such as sauerkraut contribute to the creation of a healthy gut flora, which research shows may help reduce stress and maintain brain health.

Probiotics have been found to help improve memory and lower symptoms of anxiety, depression, autism and even obsessive-compulsive disorder (OCD).

Sauerkraut may also maintain brain health by increasing your gut's absorption of mood-regulating minerals, including magnesium and zinc.

That said, some researchers warn that compounds in sauerkraut may interact with monoamine oxidase inhibitors (MAOIs), a type of medication prescribed to treat depression, anxiety disorders and Parkinson's disease.

Individuals taking these medications should check with their doctor before adding sauerkraut to their diet.

May Reduce the Risk of Certain Cancers

Cabbage, the main ingredient in sauerkraut, contains antioxidants and other beneficial plant compounds that may help reduce the risk of certain cancers.

Researchers believe these compounds may help reduce DNA damage, prevent cell mutations and block the excessive cell growth that typically leads to tumor development.

The cabbage fermentation process may also create particular plant compounds thought to help destroy precancerous cells.

Certain carcinogens are believed to become cancer-causing after they've been "activated" by particular enzymes.

Two recent studies suggest that cabbage and sauerkraut juice may help prevent cancer from developing by reducing the production of these carcinogen-activating enzymes.

In another study, researchers observed that women who ate a lot of cabbage and sauerkraut from their teens into adulthood had a reduced risk of breast cancer.

The women consuming more than three servings per week had a 72% lower risk of breast cancer than those who ate less than 1.5 servings per week.

A recent study in men shows cabbage had similar effects on the risk of prostate cancer.

However, the number of studies is limited and not all studies found the same results. Thus, more are needed before strong conclusions can be made.

May Promote Heart Health

Sauerkraut may contribute to a healthier heart.

That's because it contains a good amount of fiber and probiotics, both of which may help reduce cholesterol levels.

Probiotics such as those found in sauerkraut may also help lower blood pressure slightly. People seem to achieve the best results when they take at least 10 million CFUs per day for longer than eight weeks.

Moreover, sauerkraut is one of the rare plant sources of menaquinone, more commonly known as vitamin K2.

Vitamin K2 is believed to help reduce heart disease by preventing calcium deposits from accumulating in the arteries.

In one study, regular intake of vitamin-K2-rich foods was linked to a 57% lower risk of dying from heart disease over the 7–10 year study period.

In another, women reduced their risk of heart disease by 9% for every 10 mcg of vitamin K2 they consumed per day.

For reference, 1 cup of sauerkraut contains about 6.6 mcg of vitamin K2.

Contributes to Stronger Bones

Sauerkraut contains vitamin K2, which plays an important role in bone health.

More specifically, vitamin K2 activates two proteins whose function is to bind calcium, the main mineral found in bones.

This is thought to contribute to stronger, healthier bones. In fact, several studies have shown that vitamin K2 may benefit bone health.

For instance, a three-year study in postmenopausal women observed that those taking vitamin K2 supplements experienced a slower age-related loss in bone mineral density.

Similarly, several other studies have reported that taking vitamin K2 supplements reduced the risk of spine, hip and non-spine fractures by 60–81%.

However, some of these studies used supplements to provide very high doses of vitamin K2. Thus, it's unknown whether the vitamin K2 you'd get from eating sauerkraut alone would provide the same benefits.

Dairy Based Foods

Dairy product, milk and any of the foods made from milk, including butter, cheese, ice cream, yogurt, and condensed and dried milk.

butter stick of butter

Milk has been used by humans since the beginning of recorded time to provide both fresh and storable nutritious foods. In some countries almost half the milk produced is consumed as fresh pasteurized whole, low-fat, or skim milk. However, most milk is manufactured into more stable dairy products of worldwide commerce, such as butter, cheese, dried milks, ice cream, and condensed milk.

Cow milk (bovine species) is by far the principal type used throughout the world. Other animals utilized for their milk production include buffalo (in India, China, Egypt, and the Philippines), goats (in the Mediterranean countries), reindeer (in northern Europe), and sheep (in southern Europe). In general, the processing technology described for cow milk can be successfully applied to milk obtained from other species.

In the early 1800s the average dairy cow produced less than 1,500 litres of milk annually. With advances in animal nutrition and selective breeding, one cow now produces an average of 6,500 litres of milk a year, with some cows producing up to 10,000 litres. The Holstein-Friesian cow produces the greatest volume, but other breeds such as Ayrshire, Brown Swiss, Guernsey, and Jersey, while producing less milk, are known for supplying milk that contains higher fat, protein, and total solids.

Properties of Milk

Nutrient Composition

Although milk is a liquid and most often considered a drink, it contains between 12 and 13 percent total solids and perhaps should be regarded as a food. In contrast, many "solid" foods, such as tomatoes, carrots, and lettuce, contain as little as 6 percent solids.

Many factors influence the composition of milk, including breed, genetic constitution of the individual cow, age of the cow, stage of lactation, interval between milkings, and certain disease conditions. Since the last milk drawn at each milking is richest in fat, the completeness of milking also influences a sample. In general, the type of feed only slightly affects the composition of milk, but feed of poor quality or insufficient quantity causes both a low yield and a low percentage of total solids. Current feeding programs utilize computer technology to achieve the greatest efficiency from each animal.

The composition of milk varies among mammals, primarily to meet growth rates of the individual species. The proteins contained within the mother's milk are the major components contributing to the growth rate of the young animals. Human milk is relatively low in both proteins and minerals compared with that of cows and goats.

Goat milk has about the same nutrient composition as cow milk, but it differs in several characteristics. Goat milk is completely white in colour because all the beta-carotene (ingested from feed) is converted to vitamin A. The fat globules are smaller and therefore remain suspended, so the cream does not rise and mechanical homogenization is unnecessary. Goat milk curd forms into small, light flakes and is more easily digested, much like the curd formed from human milk. It is often prescribed for persons who are allergic to the proteins in cow milk and for some patients afflicted with stomach ulcers.

Sheep milk is rich in nutrients, having 18 percent total solids (5.8 percent protein and 6.5 percent fat). Reindeer milk has the highest level of nutrients, with 36.7 percent total solids (10.3 percent protein and 22 percent fat). These high-fat, high-protein milks are excellent ingredients for cheese and other manufactured dairy products.

The major components of milk are water, fat, protein, carbohydrate (lactose), and minerals (ash). However, there are numerous other highly important micronutrients such as vitamins, essential amino acids, and trace minerals. Indeed, more than 250 chemical compounds have been identified in milk. The table shows the composition of fresh fluid milk and other dairy products.

	Nutrient composition of dairy products (per 100 g)								
dairy product	**energy (kcal)**	**water (g)**	**protein (g)**	**fat (g)**	**carbohydrate (g)**	**cholesterol (mg)**	**vitamin A (IU)**	**riboflavin (mg)**	**calcium (mg)**
fresh milk									
whole	61	88	3.29	3.34	4.66	14	126	0.162	119
low-fat*	50	89	3.33	1.92	4.80	8	205	0.165	122
skim*	35	91	3.41	0.18	4.85	2	204	0.140	123
evaporated milk	134	74	6.81	7.56	10.04	29	243	0.316	261
evaporated skim milk*	78	79	7.55	0.20	11.35	4	392	0.309	290
sweetened condensed milk	321	27	7.91	8.70	54.40	34	328	0.416	284

nonfat dry milk*	358	4	35.10	0.72	52.19	18	2,370	1.744	1,231
butter	717	16	0.85	81.11	0.06	219	3,058	0.034	24
ice cream (vanilla)	201	61	3.50	11.00	23.60	44	409	0.240	128
ice milk (vanilla)	139	68	3.80	4.30	22.70	14	165	0.265	139
sherbet (orange)	138	66	1.10	2.00	30.40	5	76	0.068	54
frozen yogurt, nonfat	128	69	3.94	0.18	28.16	2	7	0.265	134
buttermilk	40	90	3.31	0.88	4.79	4	33	0.154	116
sour cream	214	71	3.16	20.96	4.27	44	790	0.149	116
yogurt, plain, low-fat	63	85	5.25	1.55	7.04	6	66	0.214	183
yogurt, fruit, low-fat	102	74	4.37	1.08	19.05	4	46	0.178	152
cheese									
blue	353	42	21.40	28.74	2.34	75	721	0.382	528
Brie	334	48	20.75	27.68	0.45	100	667	0.520	184
Cheddar	403	37	24.90	33.14	1.28	105	1,059	0.375	721
cottage	103	79	12.49	4.51	2.68	15	163	0.163	60
cream	349	54	7.55	34.87	2.66	110	1,427	0.197	80
mozzarella**	280	49	27.47	17.12	3.14	54	628	0.343	731
Parmesan, grated	456	18	41.56	30.02	3.74	79	701	0.386	1,376
Emmentaler (Swiss)	376	37	28.43	27.54	3.38	92	845		

*Fortified with vitamin A.
**Low moisture, part skim.

Types of Dairy Products

A dairy farm

A box of full cream, long-life milk, a bottle of strawberry drinking yogurt, and a carton of passionfruit yogurt

- Milk after optional homogenization, pasteurization, in several grades after standardization of the fat level, and possible addition of the bacteria *Streptococcus lactis* and *Leuconostoc citrovorum*
    - o *Crème fraîche*, slightly fermented cream
        - ◇ Clotted cream, thick, spoonable cream made by heating milk
        - ◇ Single cream, double cream and whipping cream

    - *Smetana*, Central and Eastern European variety of sour cream
  - Cultured milk resembling buttermilk, but uses different yeast and bacterial cultures
  - Kefir, fermented milk drink from the Northern Caucasus
  - *Kumis*, also known as *airag*, slightly fermented mares' milk popular in Central Asia
  - Powdered milk (or milk powder), produced by removing the water from (usually skim) milk
    - Whole milk products
    - Buttermilk products
    - Skim milk
    - Whey products
    - High milk-fat and nutritional products (for infant formulas)
    - Cultured and confectionery products
  - Condensed milk, milk which has been concentrated by evaporation, with sugar added for reduced process time and longer life in an opened can
  - *Khoa*, milk which has been completely concentrated by evaporation, used in Indian cuisine including gulab jamun, peda, etc.
  - Evaporated milk, (less concentrated than condensed) milk without added sugar
  - Ricotta, acidified whey, reduced in volume
  - Infant formula, dried milk powder with specific additives for feeding human infants
  - Baked milk, a variety of boiled milk that has been particularly popular in Russia
- Butter, mostly milk fat, produced by churning cream
  - Buttermilk, the liquid left over after producing butter from cream, often dried as livestock feed
  - *Ghee*, clarified butter, by gentle heating of butter and removal of the solid matter
  - *Smen*, a fermented, clarified butter used in Moroccan cooking
  - Anhydrous milkfat (clarified butter)

- Cheese, produced by coagulating milk, separating from whey and letting it ripen, generally with bacteria and sometimes also with certain molds
    - Curds, the soft, curdled part of milk (or skim milk) used to make cheese
    - *Paneer*
    - Whey, the liquid drained from curds and used for further processing or as a livestock feed
    - Cottage cheese
    - Quark
    - Cream cheese, produced by the addition of cream to milk and then curdled to form a rich curd or cheese
    - *Fromage frais*
- Casein are
    - Caseinates, sodium or calcium salts of casein
    - Milk protein concentrates and isolates
    - Whey protein concentrates and isolates, reduced lactose whey
    - Hydrolysates, milk treated with proteolytic enzymes to alter functionality
    - Mineral concentrates, byproduct of demineralizing whey
- Yogurt, milk fermented by *Streptococcus salivarius* ssp. *thermophilus* and *Lactobacillus delbrueckii* ssp. *bulgaricus* sometimes with additional bacteria, such as *Lactobacillus acidophilus*
    - *Doogh*
    - *Lassi*, Indian subcontinent
    - *Leben*
- Clabber, milk naturally fermented to a yogurt-like state
- Gelato, slowly frozen milk and water, lesser fat than ice cream
- Ice cream, slowly frozen cream, milk, flavors and emulsifying additives (dairy ice cream)
    - Ice milk, low-fat version of ice cream
    - Frozen custard
    - Frozen yogurt, yogurt with emulsifiers

- Other
  - o *Viili*
  - o *Kajmak*
  - o *Filmjölk*
  - o *Piimä*
  - o *Vla*
  - o *Dulce de leche*
  - o *Skyr*
  - o Junket, milk solidified with rennet
  - o Quiche
  - o Milk Chocolate

Consumption Patterns Worldwide

Rates of dairy consumption vary widely worldwide. High-consumption countries consume more than 150 kg per capita per year: Argentina, Armenia, Australia, Costa Rica, most European counties, Israel, Kyrgyzstan, North America and Pakistan. Medium-consumption countries consume 30 to 150 kg per capita per year: India, Iran, Japan, Kenya, Mexico, Mongolia, New Zealand, North and Southern Africa, most of the Middle East, and most of Latin America and the Caribbean. Low-consumption countries consume under 30 kg per capita per year: Senegal, most of Central Africa, and most of East and Southeast Asia.

Benefits of Milk and other Dairy Products

Milk and dairy products are the major food source of calcium and protein in most developed countries, including Australia. One litre of milk can provide approximately 1200 mg of calcium, representing more than the daily requirement for calcium. Calcium in milk is in a bioavailable form and is readily absorbed. The absorption of calcium is enhanced by vitamin A and lactose.

Adequate intake of calcium is necessary for growth in childhood, and for the prevention of diseases like osteoporosis in adulthood. Research evidence points to an association between decreased calcium intake and osteoporosis.

Calcium is also needed in the diet to control hypertension. A number of studies have reported an inverse association between dietary calcium intake and blood pressure. The most powerful research strategy is the randomised controlled trial. Findings from such clinical trials indicate that an increased calcium intake lowers blood pressure and the risk of hypertension. However, another trial in which researchers followed 193 men and women split into two groups, one taking a placebo and the other taking calcium carbonate tablets twice daily, reported that calcium had no significant effect on blood pressure.

Despite its nutritional value, some people are sceptical of milk consumption, mainly because of the fat and cholesterol content, and the risk of atherosclerosis (clogging of the arteries) or coronary heart disease (CHD). However, there is insufficient scientific evidence to support this caution. In 2004, Dairy Australia Seminar Series, a review of the epidemiological evidence on milk and cardiovascular disease presented by Professor Peter C Elwood, showed that milk drinkers have a reduced risk of cardiovascular disease (CVD) compared to those who drink little or no milk, despite the fact that most of the milk drank was regular fat milk. The review further concluded that there was no evidence of an increase in vascular disease; rather, the incidence of stroke and heart disease was about 15% lower in the subjects with the highest intakes of milk. Evidence from cohort studies suggests that though milk consumption raises cholesterol level, milk is protective against vascular disease.

The biggest setback to using milk and dairy products is that the public in general, and teenagers and women in particular, tend to avoid milk and dairy products because of their fat content.

A 300 ml glass of full cream milk contains 10 g of fat, of which 6.3 g is saturated fat, and can contribute about 770 kJ of energy. In contrast, a 300 ml glass of skim milk only contains 0.6 g of fat (a reduction of 94%), 0.3 g of saturated fat, and 440 kJ (a reduction of 43%). Therefore, consuming skim milk and low fat yoghurt instead of whole milk will allay the fear of health risks associated with dairy fat.

It has been documented that, in most countries, milk consumption has fallen over the past 20–25 years. A reduction in milk consumption may impact on nutritional needs, especially of calcium. It appears that the positive nutritional and health contributing values of dairy products far outweigh concerns regarding cholesterol.

Like any other food, overconsumption may be associated with health risks. The key word in the choice and consumption of any food is moderation. At moderate levels, calcium and dairy products have benefits beyond bone health, including possibly lowering the risk of high blood pressure and colon cancer.

Amount of Milk Needed

A daily intake of three glasses of skim milk, over cereal at breakfast, in hot beverages and in milk dishes like custard, and one or more portions of low fat yoghurt or low fat cottage cheese, is sufficient to meet your milk requirements. The US Department of Agriculture's Food Guide Pyramid recommends that individuals two years and older eat 2–3 servings of dairy products per day.

A Serving is Equal to:

- 1 cup (8 fl oz; 250 ml) of milk
- 8 oz (250 g) of yoghurt
- 1.5 oz (40 g) of natural cheese (e.g. cheddar)
- 2.0 oz (60 g) of processed cheese

Yoghurt

Yogurt is a dairy product, which is made by blending fermented milk with various ingredients that provide flavor and color. Although accidentally invented thousands of years ago, yogurt has only recently gained popularity in the United States.

It is believed that yogurt originated in Mesopotamia thousands of years ago. Evidence has shown that these people had domesticated goats and sheep around 5000 B.C. The milk from these animals was stored in gourds, and in the warm climate it naturally formed a curd. This curd was an early form of yogurt. Eventually, a process for purposely producing yogurt was developed.

While yogurt has been around for many years, it is only recently (within the last 30-40 years) that it has become popular. This is due to many factors including the introduction of fruit and other flavorings into yogurt, the convenience of it as a ready-made break-fast food and the image of yogurt as a low fat healthy food.

Manufacturers have responded to the growth in the yogurt market by introducing many different types of yogurt including low fat and no-fat, creamy, drinking, bio-yogurt, organic, baby, and frozen. Traditional yogurt is thick and creamy. It is sold plain and in a wide assortment of flavors. These are typically fruit flavors such as strawberry or blue-berry however, newer, more unique flavors such as cream pie and chocolate have also been introduced. Cereals and nuts are some-times added to yogurts. Yogurt makers also sell products with a varying level of fat. Low fat yogurt, which contains between 0.5% and 4% fat, is currently the best selling. Diet no-fat yogurt contains no fat at all. It also contains artificial sweeteners that provide sweetness while still reducing calories. Creamy yogurt is extra thick, made with whole milk and added cream. Drinking yogurt is a thinner product, which has a lower solids level than typical yogurt. Bio-yogurt is made with a different type of fermentation culture and is said to aid digestion. Yogurt that is made with milk from specially fed cows is called organic yogurt. This type of yogurt is claimed to be more nutritious than other yogurts. Other types of yogurts include pasteurized stirred yogurt that has extended shelf life, baby yogurt made specifically for children, and frozen yogurt.

The yogurt itself has a generally aldehydic flavor, which is a result of the fermentation process. Since it is made from milk, yogurt is rich in nutrients. It contains protein and vitamins and is a rich source of calcium. In fact, a small container of yogurt contains as much calcium as a third of a pint of milk. In addition to these nutritional characteristics, yogurt is also thought to have additional health benefits. One of the suggested benefits of yogurt is that it acts as a digestive aid. In the body, it is thought that yogurt can encourage the growth of beneficial bacteria in the gut. These organisms help to digest food more efficiently and protect against other, harmful organisms. Another health benefit of yogurt is for people that are lactose intolerant. These people have difficulty digesting milk products however, they typically can tolerate yogurt.

Raw Materials

In general, yogurt is made with a variety of ingredients including milk, sugars, stabilizers, fruits and flavors, and a bacterial culture

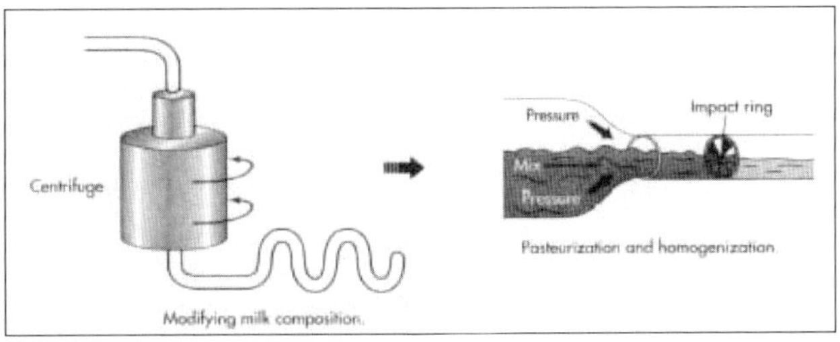

When the milk arrives at the plant, its composition is modified before it is used to make yogurt.

This standardization process typically involves reducing the fat content and increasing the total solids. Once modification occurs, it is pasteurized to kill bacteria and homogenized to consistently disperse fat molecules.

(Lactobacillus bulgaricus). During fermentation, these organisms interact with the milk and convert it into a curd. They also change the flavor of the milk giving it the characteristic yogurt flavor of which acetaldehyde is one of the important contributors. The primary byproduct of the fermentation process is lactic acid. The acid level is used to determine when the yogurt fermentation is completed which is usually three to four hours. The suppliers of these yogurt cultures offer various combinations of the two bacterial types to produce yogurts with different flavors and textures.

To modify certain properties of the yogurt, various ingredients may be added. To make yogurt sweeter, sucrose (sugar) may be added at approximately 7%. For reduced calorie yogurts, artificial sweeteners such as aspartame or saccharin are used. Cream may be added to provide a smoother texture. The consistency and shelf stability of the yogurt can be improved by the inclusion of stabilizers such as food starch, gelatin, locust-bean gum, guar gum and pectin. These materials are used because they do not have a significant impact on the final flavor. The use of stabilizers is not required however, and some marketers choose not to use them in order to retain a more natural image for their yogurt.

To improve taste and provide a variety of flavors, many kinds of fruits are added to yogurt. Popular fruits include strawberries, blueberries, bananas, and peaches, but almost any fruit can be added. Beyond fruits, other flavorings are also added. These can include such things as vanilla, chocolate, coffee, and even mint. Recently, manufacturers have become quite creative in the types of yogurt they produce using natural and artificial flavorings.

The Manufacturing Process

The general process of making yogurt includes modifying the composition of and pasteurizing the milk; fermenting at warm temperatures; cooling it; and adding fruit, sugar, and other materials.

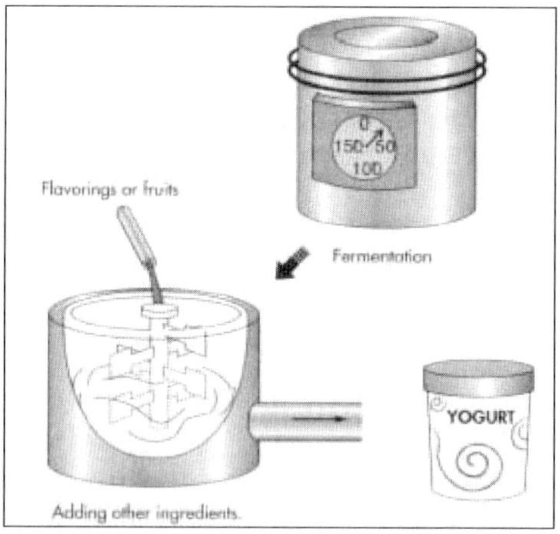

The milk substance is fermented until it becomes yogurt. Fruits and flavorings are added to the yogurt before packaging.

Modifying milk composition

- When the milk arrives at the plant, its composition is modified before it is used to make yogurt. This standardization process typically involves reducing the fat content and increasing the total solids. The fat content is reduced by using a standardizing clarifier and a separator (a device that relies upon centrifugation to separate fat from milk). From the clarifier, the milk is placed in a storage tank and tested for fat and solids content. For yogurt manufacture, the solids content of the milk is increased to 16% with 1-5% being fat and 11-14% being solids-not-fat (SNF). This is accomplished either by evaporating off some of the water, or adding concentrated milk or milk powder. Increasing the solids content improves the nutritional value of the yogurt, makes it easier to produce a firmer yogurt and improves the stability of the yogurt by reducing the tendency for it to separate on storage.

Pasteurization and homogenization

- After the solids composition is adjusted, stabilizers are added and the milk is pasteurized. This step has many benefits. First, it will destroy all the microorganisms in the milk that may interfere with the controlled fermentation process. Second, it will denature the whey proteins in the milk which will give the final yogurt product better body and texture. Third, it will not greatly alter the flavor of the milk. Finally, it helps release the compounds in milk that will stimulate the growth of the starter culture. Pasteurization can be a continuous-or batch-process. Both of these processes involve heating the milk to a relatively high temperature and holding it there for a set amount of time. One specific method for batch process pasteurization is to heat a large, stainless steel vat of milk to 185° F (85° C) and hold it there for at least 30 minutes.

- While the milk is being heat treated, it is also homogenized. Homogenization is a process in which the fat globules in milk are broken up into smaller, more consistently dispersed particles. This produces a much smoother and creamier end product. In commercial yogurt making, homogenization has the benefits of giving a uniform product, which will not separate. Homogenization is accomplished using a homogenizer or viscolizer. In this machine, the milk is forced through small openings at a high pressure and fat globules are broken up due to shearing forces.

Fermentation

- When pasteurization and homogenization are complete, the milk is cooled to between 109.4-114.8° F (43-46° C) and the fermentation culture is added in a concentration of about 2%. It is held at this temperature for about three to four hours while the incubation process takes place. During this time, the bacteria metabolizes certain compounds in the milk producing the characteristic yogurt flavor. An important byproduct of this process is lactic acid.

- Depending on the type of yogurt, the incubation process is done either in a large tank of several hundred gallons or in the final individual containers. Stirred yogurt is fermented in bulk and then poured into the final selling containers. Set yogurt, also known as French style, is allowed to ferment right in the container it is sold in. In both instances, the lactic acid level is used to determine when the yogurt is ready. The acid level is found by taking a sample of the product and titrating it with sodium hydroxide. A value of at least 0.9% acidity and a pH of about 4.4 are the current minimum standards for yogurt manufacture in the United States. When the yogurt reaches the desired acid level, it is cooled, modified as necessary and dispensed into containers (if applicable).

Adding other ingredients

- Fruits, flavors, and other additives can be added to the yogurt at various points in manufacturing process. This is typically dependent on the type of yogurt being produced. Flavor in non-fruit yogurts are added to the process milk before being dispensed into cartons. Fruits and flavors can also be added to the containers first, creating a bottom layer. The inoculated milk is then added on top and the carton is sealed and incubated. If the fruit is pasteurized, it can be added as a puree to the bulk yogurt, which is then dispensed into containers. Finally, the fruit can be put into a special package, which is mixed with plain yogurt upon consumption.
- The finished yogurt containers are placed in cardboard cases, stacked on pallets, and delivered to stores via refrigerated trucks.

Quality Control

Milk products such as yogurt are subject to a variety of safety testing. Some of these include tests for microbial quality, degree of pasteurization, and various forms of contaminants. The microbial quality of the incoming milk is determined by using a dye reaction test. This method shows the number of organisms present in the incoming milk. If the microbial count is too high at this point, the milk may not be used for manufacture. Since complete pasteurization inactivates most organisms in milk, the degree of pasteurization is determined by measuring the level of an enzyme in the milk called phosphatase. Governmental regulations require that this test be run to ensure that pasteurization is done properly. Beyond microbial contamination, raw milk is subject to other kinds of contaminants such as antibiotics, pesticides or even radioactivity. These can all be found through safety testing and the milk is treated accordingly.

In addition to safety tests, the final yogurt product is also evaluated to ensure that it meets the specifications set by the manufacturer for characteristics such as pH, rheology, taste, color, and odor. These factors are tested using various laboratory equipment such as pH meters and viscometers and also human panelists.

Cheese

Cheese is a generic term for a diverse group of milk-based food products. Cheese is produced in wide-ranging flavors, textures, and forms.

Cheese consists of proteins and fat from milk, usually the milk of cows, buffalo, goats, or sheep. It is produced by coagulation of the milk protein casein. Typically, the milk is acidified and addition of the enzyme rennet causes coagulation. The solids are separated and pressed into final form. Some cheeses have molds on the rind or throughout. Most cheeses melt at cooking temperature.

Hundreds of types of cheese from various countries are produced. Their styles, textures and flavors depend on the origin of the milk (including the animal's diet), whether they have been pasteurized, the butterfat content, the bacteria and mold, the processing, and aging. Herbs, spices, or wood smoke may be used as flavoring agents. The yellow to red color of many cheeses, such as Red Leicester, is normally formed from adding annatto.

For a few cheeses, the milk is curdled by adding acids such as vinegar or lemon juice. Most cheeses are acidified to a lesser degree by bacteria, which turn milk sugars into lactic acid, then the addition of rennet completes the curdling. Vegetarian alternatives to rennet are available; most are produced by fermentation of the fungus Mucor miehei, but others have been extracted from various species of the Cynara thistle family.

Cheese is valued for its portability, long life, and high content of fat, protein, calcium, and phosphorus. Cheese is more compact and has a longer shelf life than milk, although how long a cheese will keep may depend on the type of cheese; labels on packets of cheese often claim that a cheese should be consumed within three to five days of opening. Generally speaking, hard cheeses last longer than soft cheeses, such as Brie or goat's milk cheese. Cheesemakers near a dairy region may benefit from fresher, lower-priced milk, and lower shipping costs. The long storage life of some cheese, especially if it is encased in a protective rind, allows selling when markets are favorable.

Production

French Cheeses

In 2014, world production of cheese from whole cow milk was 18.7 million tonnes, with the United States accounting for 29% (5.4 million tonnes) of the world total followed by Germany, France and Italy as major producers.

Other 2014 world totals for processed cheese include:

- from skimmed cow milk, 2.4 million tonnes (leading country, Germany, 845,500 tonnes)
- from goat milk, 523,040 tonnes (leading country, South Sudan, 110,750 tonnes)
- from sheep milk, 680,302 tonnes (leading country, Greece, 125,000 tonnes)
- from buffalo milk, 282,127 tonnes (leading country, Egypt, 254,000 tonnes)

During 2015, Germany, France, Netherlands and Italy exported 10-14% of their produced cheese. The United States was a marginal exporter (5.3% of total cow milk production), as most of its output was for the domestic market.

Consumption

France, Iceland, Finland, Denmark and Germany were the highest consumers of cheese in 2014, averaging 25 kg (55 lb) per person.

Processing

Curdling

A required step in cheesemaking is separating the milk into solid curds and liquid whey. Usually this is done by acidifying (souring) the milk and adding rennet. The acidification can be accomplished directly by the addition of an acid, such as vinegar, in a few cases (paneer, queso fresco). More commonly starter bacteria are employed instead which convert milk sugars into lactic acid. The same bacteria (and the enzymes they produce) also play a large role in the eventual flavor of aged cheeses. Most cheeses are made with starter bacteria from the *Lactococcus*, *Lactobacillus*, or *Streptococcus* families. Swiss starter cultures also include *Propionibacter shermani*, which produces carbon dioxide gas bubbles during aging, giving Swiss cheese or Emmental its holes (called "eyes").

Some fresh cheeses are curdled only by acidity, but most cheeses also use rennet. Rennet sets the cheese into a strong and rubbery gel compared to the fragile curds produced by acidic coagulation alone. It also allows curdling at a lower acidity—important because flavor-making bacteria are inhibited in high-acidity environments. In general, softer, smaller, fresher cheeses are curdled with a greater proportion of acid to rennet than harder, larger, longer-aged varieties.

While rennet was traditionally produced via extraction from the inner mucosa of the fourth stomach chamber of slaughtered young, unweaned calves, most rennet used today in cheesemaking is produced recombinantly. The majority of the applied chymosin is retained in the whey and, at most, may be present in cheese in trace quantities. In ripe cheese, the type and provenance of chymosin used in production cannot be determined.

During industrial production of Emmental cheese, the as-yet-undrained curd is broken by rotating mixers.

Curd Processing

At this point, the cheese has set into a very moist gel. Some soft cheeses are now essentially complete: they are drained, salted, and packaged. For most of the rest, the curd is cut into small cubes. This allows water to drain from the individual pieces of curd.

Some hard cheeses are then heated to temperatures in the range of 35–55 °C (95–131 °F). This forces more whey from the cut curd. It also changes the taste of the finished cheese, affecting both the bacterial culture and the milk chemistry. Cheeses that are heated to the higher temperatures are usually made with thermophilic starter bacteria that survive this step—either *Lactobacilli* or *Streptococci*.

Salt has roles in cheese besides adding a salty flavor. It preserves cheese from spoiling, draws moisture from the curd, and firms cheese's texture in an interaction with its proteins. Some cheeses are salted from the outside with dry salt or brine washes. Most cheeses have the salt mixed directly into the curds.

Cheese factory in the Netherlands

Other techniques influence a cheese's texture and flavor. Some examples are :

- Stretching: (Mozzarella, Provolone) The curd is stretched and kneaded in hot water, developing a stringy, fibrous body.
- Cheddaring: (Cheddar, other English cheeses) The cut curd is repeatedly piled up, pushing more moisture away. The curd is also mixed (or *milled*) for a long time, taking the sharp edges off the cut curd pieces and influencing the final product's texture.
- Washing: (Edam, Gouda, Colby) The curd is washed in warm water, lowering its acidity and making for a milder-tasting cheese.

Most cheeses achieve their final shape when the curds are pressed into a mold or form. The harder the cheese, the more pressure is applied. The pressure drives out moisture—the molds are designed to allow water to escape—and unifies the curds into a single solid body.

Parmigiano-Reggiano in a modern factory

Ripening

A newborn cheese is usually salty yet bland in flavor and, for harder varieties, rubbery in texture. These qualities are sometimes enjoyed—cheese curds are eaten on their own—but normally cheeses are left to rest under controlled conditions. This aging period (also called ripening, or, from the French, *affinage*) lasts from a few days to several years. As a cheese ages, microbes and enzymes transform texture and intensify flavor. This transformation is largely a result of the breakdown of casein proteins and milkfat into a complex mix of amino acids, amines, and fatty acids.

Some cheeses have additional bacteria or molds intentionally introduced before or during aging. In traditional cheesemaking, these microbes might be already present in the aging room; they are simply allowed to settle and grow on the stored cheeses. More often today, prepared cultures are used, giving more consistent results and putting fewer constraints on the environment where the cheese ages. These cheeses include soft ripened cheeses such as Brie and Camembert, blue cheeses such as Roquefort, Stilton, Gorgonzola, and rind-washed cheeses such as Limburger.

Types

Crottin Cheese

Lagrein Weinkäse cheese

Murol cheese

There are many types of cheese, with around 500 different varieties recognized by the

International Dairy Federation, more than 400 identified by Walter and Hargrove, more than 500 by Burkhalter, and more than 1,000 by Sandine and Elliker. The varieties may be grouped or classified into types according to criteria such as length of ageing, texture, methods of making, fat content, animal milk, country or region of origin, etc.—with these criteria either being used singly or in combination, but with no single method being universally used. The method most commonly and traditionally used is based on moisture content, which is then further discriminated by fat content and curing or ripening methods. Some attempts have been made to rationalise the classification of cheese—a scheme was proposed by Pieter Walstra which uses the primary and secondary starter combined with moisture content, and Walter and Hargrove suggested classifying by production methods which produces 18 types, which are then further grouped by moisture content.

Moisture Content (Soft to Hard)

Categorizing cheeses by firmness is a common but inexact practice. The lines between "soft", "semi-soft", "semi-hard", and "hard" are arbitrary, and many types of cheese are made in softer or firmer variations. The main factor that controls cheese hardness is moisture content, which depends largely on the pressure with which it is packed into molds, and on aging time.

Fresh, Whey and Stretched Curd Cheeses

The main factor in the categorization of these cheeses is their age. Fresh cheeses without additional preservatives can spoil in a matter of days.

Content (Double Cream, Goat, Ewe and Water Buffalo)

Some cheeses are categorized by the source of the milk used to produce them or by the added fat content of the milk from which they are produced. While most of the world's commercially available cheese is made from cows' milk, many parts of the world also produce cheese from goats and sheep. Double cream cheeses are soft cheeses of cows' milk enriched with cream so that their fat content is 60% or, in the case of triple creams, 75%. The use of the terms "double" or "triple" is not meant to give a quantitative reference to the change in fat content, since the fat content of whole cows' milk is 3%-4%.

Soft-ripened and Blue-vein

There are at least three main categories of cheese in which the presence of mold is a significant feature: soft ripened cheeses, washed rind cheeses and blue cheeses.

Processed Cheeses

Processed cheese is made from traditional cheese and emulsifying salts, often with the addition of milk, more salt, preservatives, and food coloring. It is inexpensive, consistent, and melts smoothly. It is sold packaged and either pre-sliced or unsliced, in a number of varieties. It is also available in aerosol cans in some countries.

Cooking and Eating

At refrigerator temperatures, the fat in a piece of cheese is as hard as unsoftened butter, and its protein structure is stiff as well. Flavor and odor compounds are less easily liberated when cold. For improvements in flavor and texture, it is widely advised that cheeses be allowed to warm up to room temperature before eating. If the cheese is further warmed, to 26–32 °C (79–90 °F), the fats will begin to "sweat out" as they go beyond soft to fully liquid.

Above room temperatures, most hard cheeses melt. Rennet-curdled cheeses have a gel-like protein matrix that is broken down by heat. When enough protein bonds are broken, the cheese itself turns from a solid to a viscous liquid. Soft, high-moisture cheeses will melt at around 55 °C (131 °F), while hard, low-moisture cheeses such as Parmesan remain solid until they reach about 82 °C (180 °F). Acid-set cheeses, including halloumi, paneer, some whey cheeses and many varieties of fresh goat cheese, have a protein structure that remains intact at high temperatures. When cooked, these cheeses just get firmer as water evaporates.

Some cheeses, like raclette, melt smoothly; many tend to become stringy or suffer from a separation of their fats. Many of these can be coaxed into melting smoothly in the presence of acids or starch. Fondue, with wine providing the acidity, is a good example of a smoothly melted cheese dish. Elastic stringiness is a quality that is sometimes enjoyed, in dishes including pizza and Welsh rarebit. Even a melted cheese eventually turns solid again, after enough moisture is cooked off. The saying "you can't melt cheese twice" (meaning "some things can only be done once") refers to the fact that oils leach out during the first melting and are gone, leaving the non-meltable solids behind.

Zigerbrüt, cheese grated onto bread through a mill, from the Canton of Glarus in Switzerland.

As its temperature continues to rise, cheese will brown and eventually burn. Browned, partially burned cheese has a particular distinct flavor of its own and is frequently used in cooking (e.g., sprinkling atop items before baking them).

Cheeseboard

Various cheeses on a cheeseboard served with wine for lunch

A cheeseboard (or cheese course) may be served at the end of a meal, either replacing or following dessert. A cheeseboard typically has contrasting cheeses with accompaniments, such as crackers, biscuits, grapes, nuts, celery or chutney. Port or other dessert wines may be served with a cheeseboard. A cheeseboard 70 feet (21 m) long was used to feature the variety of cheeses manufactured in Wisconsin, where the state legislature recognizes a "cheesehead" hat as a state symbol.

Nutrition and Health

The nutritional value of cheese varies widely. Cottage cheese may consist of 4% fat and 11% protein while some whey cheeses are 15% fat and 11% protein, and triple-crème cheeses are 36% fat and 7% protein. In general, cheese is a rich source (20% or more of the Daily Value, DV) of calcium, protein, phosphorus, sodium and saturated fat. A 28-gram (one ounce) serving of cheddar cheese contains about 7 grams (0.25 oz) of protein and 202 milligrams of calcium. Nutritionally, cheese is essentially concentrated milk: it takes about 200 grams (7.1 oz) of milk to provide that much protein, and 150 grams (5.3 oz) to equal the calcium.

MacroNutrients (grams) of common cheeses per 100gm

Cheese	Water	Protein	Fat	Carbs
Swiss	37.1	26.9	27.8	5.4
Feta	55.2	14.2	21.3	4.1
Cheddar	36.8	24.9	33.1	1.3
Mozzarella	50	22.2	22.4	2.2
Cottage	80	11.1	4.3	3.4

Vitamin contents in %DV of common cheeses per 100gm

Cheese	A	B1	B2	B3	B5	B6	B9	B12	Ch.	C	D	E	K
Swiss	17	4	17	0	4	4	1	56	2.8	0	11	2	3
Feta	8	10	50	5	10	21	8	28	2.2	0	0	1	2
Cheddar	20	2	22	0	4	4	5	14	3	0	3	1	3
Mozzarella	14	2	17	1	1	2	2	38	2.8	0	0	1	3
Cottage	3	2	10	0	6	2	3	7	3.3	0	0	0	0

Mineral contents in %DV of common cheeses per 100 grams

Cheese	Ca	Fe	Mg	P	K	Na	Zn	Cu	Mn	Se
Swiss	79	10	1	57	2	8	29	2	0	26
Feta	49	4	5	34	2	46	19	2	1	21
Cheddar	72	4	7	51	3	26	21	2	1	20
Mozzarella	51	2	5	35	2	26	19	1	1	24
Cottage	8	0	2	16	3	15	3	1	0	14

Ch. = Choline; Ca = Calcium; Fe = Iron; Mg = Magnesium; P = Phosphorus; K = Potassium; Na = Sodium; Zn = Zinc; Cu = Copper; Mn = Manganese; Se = Selenium;

Neonatal Infection and Death

Cheese has the potential for promoting the growth of *Listeria* bacteria. *Listeria monocytogenes* can also cause serious infection in an infant and pregnant woman and can be transmitted to her infant in utero or after birth. The infection has the potential of seriously harming or even causing the death of a preterm infant, an infant of low or very low birth weight, or an infant with an immune system deficiency or a congenital defect of the immune system. The presence of this pathogen can sometimes be determined by the symptoms that appear as a gastrointestinal illness in the mother. The mother can also acquire infection from ingesting food that contains other animal products such as, unpasteurized milk, delicatessen meats, and hot dogs.

Heart Disease

A review of the medical literature published in 2012 noted that: "Cheese consumption is the leading contributor of SF (saturated fat) in the U.S. diet, and therefore would be predicted to increase LDL-C (LDL cholesterol) and consequently increase the risk of CVD (cardiovascular disease)." It found that: "Based on results from numerous prospective observational studies and meta-analyses, most, but not all, have shown no association and in some cases an inverse relationship between the intake of milk fat containing dairy products and the risk of CVD, CHD (coronary heart disease), and stroke. A limited number of prospective cohort studies found no significant association between the intake of total full-fat dairy products and the risk of CHD or stroke. Most clinical studies showed that full-fat natural cheese, a highly fermented product, significantly lowers LDL-C compared with butter intake of equal total fat and saturated fat content."

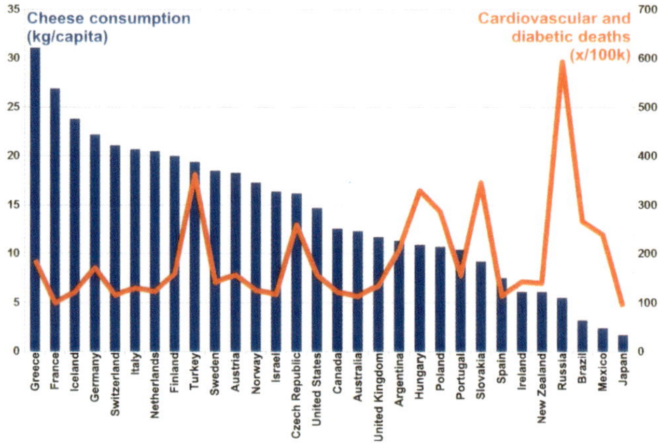

Average cheese consumption and rates of mortality due to cardiovascular disease or diabetes

Pasteurization

A number of food safety agencies around the world have warned of the risks of raw-milk cheeses. The U.S. Food and Drug Administration states that soft raw-milk cheeses can cause "serious infectious diseases including listeriosis, brucellosis, salmonellosis and tuberculosis". It is U.S. law since 1944 that all raw-milk cheeses (including imports since 1951) must be aged at least 60 days. Australia has a wide ban on raw-milk cheeses as well, though in recent years exceptions have been made for Swiss Gruyère, Emmental and Sbrinz, and for French Roquefort. There is a trend for cheeses to be pasteurized even when not required by law.

Pregnant women may face an additional risk from cheese; the U.S. Centers for Disease Control has warned pregnant women against eating soft-ripened cheeses and blue-veined cheeses, due to the listeria risk, which can cause miscarriage or harm the fetus.

Buttermilk

What is Buttermilk? There's the old answer and the new, and both are good.

The old answer is that buttermilk was the thin, non-fat but rich tasting liquid left in a churn after making butter, full of healthful, delicious cultures that develop naturally when cream is left at room temperature for a few hours to improve the flavor of the butter. The cultures meant that buttermilk kept longer than raw milk in the days before easy cooling and refrigeration, which made it useful in cooking.

The new answer is that buttermilk is still cultured milk, similar to natural yogurt and kefir, but instead of being a by-product of churning most dairies inoculate fresh, pasteurized milk with cultures (harmless lactic acid bacteria) that transform it into the buttermilk we buy in bottles and cartons in stores.

Although it looks and tastes rich and creamy, traditional churned buttermilk was always nonfat because all the fat wound up in the homemade butter. These days, cultured buttermilk can range from skim to full-fat with corresponding calorie count, just like yogurt and sour cream, although most of what we buy in stores is low-fat.

Store-bought buttermilk is thicker, tangier, and more acidic than traditional or homemade buttermilk, so if you're preparing a recipe that calls for buttermilk, it's best to stick with store-bought, especially in baked goods that depend on precise leavening. Many recipes that call for buttermilk include baking soda as part of the leavening, to balance the acidity in the commercial buttermilk.

Buttermilk is a prized Southern ingredient, but it's also a delicious beverage that's good for us. It is a potent source of probiotics and active cultures found in natural yogurt, kefir, kombucha, sauerkraut, kimchee, and other fermented foods that aid digestion and promote gut health. A glass of bedtime buttermilk has soothed the tummies of the dyspeptic and the frayed nerves of many a late night reveler. A few sips calmed some things down and perked other things up.

Traditional Buttermilk

Originally, buttermilk referred to the liquid left over from churning butter from cultured or fermented cream. Traditionally, before cream could be skimmed from whole milk, the milk was left to sit for a period of time to allow the cream and milk to separate. During this time, naturally occurring lactic acid-producing bacteria in the milk fermented it. This facilitates the butter churning process, since fat from cream with a lower pH coalesces more readily than that of fresh cream. The acidic environment also helps prevent potentially harmful microorganisms from growing, increasing shelf-life. However, in establishments that used cream separators, the cream is hardly acidic at all.

On the Indian subcontinent, the term "buttermilk" refers to the liquid left over after extracting butter from churned cream. Today, this is called *traditional buttermilk*. Traditional buttermilk is still common in many Indian, Nepalese, and Pakistani households, but rarely found in Western countries. In Nepal, buttermilk is called *mohi* and is a common drink in many Nepalese homes. It is served to family members and guests, and can be taken with meals or snacks. In many families, it is most popularly served with roasted maize.

Buttermilk can be made at home by whisking a cup of non-homogenized cream until it separates into butter and buttermilk.

Cultured Buttermilk

Commercially available cultured buttermilk is milk that has been pasteurized and homogenized (with 1% or 2% fat), and then inoculated with a culture of *Lactococcus lactis* plus *Leuconostoc citrovorum* to simulate the naturally occurring bacteria in the old-fashioned product. Some dairies add colored flecks of butter to cultured buttermilk to simulate residual flecks of butter that can be left over from the churning process of traditional buttermilk.

Cultured buttermilk, often known simply as "buttermilk" now, was first commercially introduced in the United States in the 1920s. It was popular among immigrants, and viewed as a food that could slow aging. It reached peak annual sales of 517,000,000 kilograms (1.140×10^9 lb) in 1960. Buttermilk's popularity has declined since then, despite an increasing population, and annual sales in 2012 reached less than half that number.

Condensed buttermilk and dried buttermilk have increased in importance in the food industry. Buttermilk solids are used in ice cream manufacturing, as well as being added to pancake mixes to make buttermilk pancakes.

Acidified Buttermilk

Acidified buttermilk is a related product made by adding a food-grade acid (such as lemon juice) to milk. It can be produced by mixing 1 tablespoon of vinegar or lemon juice per 1 cup of milk and letting it sit until it curdles, about 10 minutes. Any level of fat content for the milk ingredient may be used, but whole milk is usually used for baking. In the process which is used to produce paneer, such acidification is done in the presence of heat.

Powdered Buttermilk

Like powdered milk, buttermilk is available in a dried powder form. This stores well at room temperature and is usually used in baked goods.

Nutrition

One cup (237 ml) of whole milk contains 157 calories and 8.9 grams of fat whereas one cup of buttermilk contains 99 calories and 2.2 grams of fat. Buttermilk contains vitamins, potassium, calcium, and traces of phosphorus.

Kefir

Kefir is a cultured, fermented beverage that tastes a great deal like a yogurt drink. It's made using "starter" grains, just as sourdough bread has a "starter." This starter is a combination of yeasts, milk proteins, and bacteria. It has a tart, creamy flavor, and it's loaded with probiotic health benefits.

Kefir is most commonly made with dairy milk, but it can be made with non-dairy alternatives including:

- coconut milk
- goat's milk
- rice milk
- coconut water

Because kefir is fermented, most people who are lactose intolerant can actually drink kefir.

Kefir is now widely accessible in most places. It can be found in some form in most grocery stores near the dairy or yogurt. It's often used to improve digestion.

Kefir vs. Yogurt

Kefir and yogurt are sometimes lumped together as being nearly identical products, but this isn't accurate.

Kefir and yogurt do have a lot in common. They both have similar creamy-but-tart tastes and are traditionally made from dairy (but can be made with alternatives). They also both have plenty of protein, calcium, B vitamins, potassium, and probiotics.

However, they do have significant differences. Kefir typically has more fat than yogurt, but it also has more protein and more probiotics. Kefir is also thinner and is best as a drink. Yogurt has a thicker consistency.

Kefir and yogurt are made differently. Kefir ferments at room temperature, while many types of yogurt start culturing under heat. Kefir contains a larger number of different types of healthy bacteria, and it's better for the gastrointestinal tract because of this. Kefir's active yeast also has more nutritional benefits than that of yogurt.

Traditional kefir is fermented at ambient temperatures, generally overnight. Fermentation of the lactose yields a sour, carbonated, slightly alcoholic beverage, with a consistency and taste similar to thin yogurt.

The kefir grains initiating the fermentation are a combination of lactic acid bacteria and yeasts in a matrix of proteins, lipids, and sugars. This symbiotic culture of bacteria and yeast (or SCOBY) forms "grains" that resemble cauliflower. A complex and highly variable community of lactic acid bacteria and yeasts can be found in these grains, although some predominate; *Lactobacillus* species are always present. Even successive batches of kefir may differ due to factors such as the kefir grains rising out of the milk while fermenting or curds forming around the grains, as well as room temperature.

Kefir grains contain kefiran, a water-soluble polysaccharide, which imparts a creamy texture and feeling in the mouth. The grains range in color from white (the acceptable color of healthy grains) to yellow. The latter is the outcome of leaving the grains in the same milk during fermentation for longer than the optimal 24-hour period and continually doing so over many batches. Grains may grow to the size of walnuts or even larger.

The composition of kefir depends greatly on the type of milk that was fermented, including the concentration of vitamin B_{12}.

Kefir preparation

During fermentation, changes in composition of nutrients and other ingredients occur. Lactose, the sugar present in milk, is broken down mostly to lactic acid (25%) by the lactic acid bacteria, which results in acidification of the product. Propionibacteria further break down some of the lactic acid into propionic acid (these bacteria also carry out the same fermentation in Swiss cheese). A portion of lactose is converted to kefiran, which is indigestible by gastric digestion. Other substances that contribute to the flavor of kefir are pyruvic acid, acetic acid, diacetyl and acetoin (both of which contribute a "buttery" flavor), citric acid, acetaldehyde, and amino acids resulting from protein breakdown.

The slow-acting yeasts, late in the fermentation process, break lactose down into ethanol and carbon dioxide. Depending on the process, ethanol concentration can be as high as 1–2% (achieved by small-scale dairies early in the 20th century), with the kefir having a bubbly appearance and carbonated taste. This makes kefir different from yogurt and most other sour milk products where only bacteria ferment the lactose into acids. Most modern processes, which use shorter fermentation times, result in much lower ethanol concentrations of 0.2–0.3%.

As a result of the fermentation, very little lactose remains in kefir. People with lactose intolerance are able to tolerate kefir, provided the number of live bacteria present in this beverage consumed is high enough (i.e., fermentation has proceeded for adequate time). It has also been shown that fermented milk products have a slower transit time than milk, which may further improve lactose digestion.

For the preparation of commercially-prepared kefir, so-called mild kefir, kefir grains are no longer used, but rather a powdered preparation of bacteria and yeast, allowing the flavor to be kept constant.

Variationsthat thrive in various other liquids exist, and vary markedly from kefir in both appearance and microbial composition. Water kefir (or *tibicos*) is grown for a day or more at room temperature in water with sugar, sometimes with lemon juice and added dry fruit such as figs.

Nutrition

Nutritional Composition

Kefir products contain nutrients in varying amounts from negligible to significant, including dietary minerals, vitamins, essential amino acids, and conjugated linoleic acid, in amounts similar to unfermented cow, goat, or sheep milk. At a pH of 4.2 - 4.6, kefir is composed mainly of water and by-products of the fermentation process, including carbon dioxide and ethanol.

Typical of milk, several dietary minerals are found in kefir, such as calcium, iron, phosphorus, magnesium, potassium, sodium, copper, molybdenum, manganese, and zinc in amounts that have not been standardized to a reputable nutrient database. Also similar to milk, kefir contains vitamins in variable amounts, including vitamin A, vitamin B_1 (thiamine), vitamin B_2 (riboflavin), vitamin B_3 (niacin), vitamin B_6 (pyridoxine), vitamin B_9 (folic acid), vitamin B_{12} (cyanocobalamin), vitamin C, vitamin D, and vitamin E. Essential amino acids found in kefir include methionine, cysteine, tryptophan, phenylalanine, tyrosine, leucine, isoleucine, threonine, lysine, and valine, as for any milk product.

Probiotics

Probiotic bacteria found in kefir products include *Lactobacillus acidophilus, Bifidobacterium bifidum, Streptococcus thermophilus, Lactobacillus delbrueckii subsp. bulgaricus, Lactobacillus helveticus, Lactobacillus kefiranofaciens, Lactococcus lactis,* and *Leuconostoc* species.

Lactobacilli in kefir may exist in concentrations varying from approximately 1 million to 1 billion colony-forming units per milliliter and are the bacteria responsible for the synthesis of the polysaccharide kefiran.

In addition to bacteria, kefir often contains strains of yeast that can metabolize lactose, such as *Kluyveromyces marxianus, Kluyveromyces lactis,* and *Saccharomyces fragilis,* as well as strains of yeast that do not metabolize lactose, including *Saccharomyces cerevisiae, Torulaspora delbrueckii,* and *Kazachstania unispora*. The nutritional significance of these strains is unknown.

Production

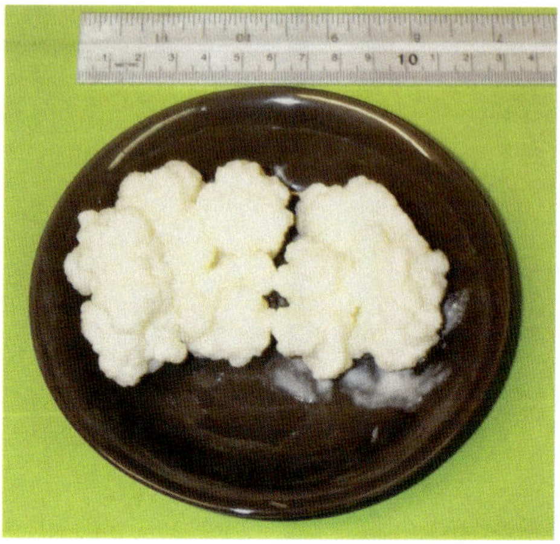

90 grams of kefir grains

Production of traditional kefir requires a starter community of kefir grains which are added to milk.

The traditional or artisanal method of making kefir is to add 2–10% kefir grains directly to milk in a sealed goatskin leather bag, which is traditionally agitated one or more times a day. Today the leather bag is replaced with a corrosion-resistant container such as a glass jar. It is not filled to capacity, allowing room for some expansion as the carbon dioxide gas produced causes the liquid level to rise. Non-lightproof containers are stored in the dark to prevent degradation of light-sensitive vitamins. After a period of fermentation lasting around 24 hours, ideally at 20–25 °C (68–77 °F), the grains are strained from the liquid using a corrosion-resistant (stainless steel or plastic) utensil and kept as the starter for another batch.

The fermented liquid-kefir, which contains live micro-organisms from the grains, may be drunk, used in recipes, or kept aside in a sealed container for many days to undergo a slower secondary fermentation in which bio-synthesis by certain micro-organisms increases the content of folic acid and some other B vitamins, and the kefir becomes sourer. The shelf life, unrefrigerated, is up to thirty days. The grains enlarge in the process of kefir production and eventually split.

The Russian method permits production of kefir on a larger scale and uses two fermentations. The first step is to prepare the cultures by incubating milk with 2–3% grains as described. The grains are then removed by filtration and 1–3% of the resulting liquid mother culture is added to milk and fermented for 12 to 18 hours.

Kefir can be made using freeze-dried cultures commonly available in powder form from health food stores. A portion of the resulting kefir can be saved to be used a number of times to propagate further fermentations but ultimately does not form grains, so a fresh culture must be obtained.

Milk Types

Kefir grains will ferment the milk from most mammals and will continue to grow in such milk. Typical milks used include cow, goat, and sheep, each with varying organoleptic (flavor, aroma, and texture) and nutritional qualities. Raw milk has been traditionally used.

Kefir grains will also ferment milk substitutes such as soy milk, rice milk, and coconut milk, as well as other sugary liquids including fruit juice, coconut water, beer wort, and ginger beer. However, the kefir grains may cease growing if the medium used does not contain all the growth factors required by the bacteria.

Milk sugar is not essential for the synthesis of the polysaccharide that makes up the grains (kefiran), and rice hydrolysate is a suitable alternative medium. Additionally, kefir grains will reproduce when fermenting soy milk, although they will change in appearance and size due to the differing proteins available to them.

Regional Consumption

Kefir, first consumed in the Caucasus, Russia and Central Asian countries for centuries, spread to the rest of Europe, Japan, the United States by early 21st century. In Chile, where it is known as "yogur de pajaritos" (little birds' yogurt), kefir has been regularly consumed for over a century.

Culinary uses

Lithuanian kefir-based cold borscht

As it contains *Lactobacillus* bacteria, kefir can be used to make a sourdough bread. It is also useful as a buttermilk substitute in baking. Kefir is one of the main ingredients in cold borscht in Lithuania, also known in Poland as Lithuanian cold soup (chłodnik litewski), and other countries.

Kefir-based soup okroshka is popular across the former Soviet Union.

Kefir may be used in place of milk on cereal, granola or milkshakes.

Health Benefits of Kefir

Part of the reason kefir has become more accessible and widespread in the past few years is likely due to its health benefits.

Kefir is nutrient-dense, with plenty of protein, B vitamins, potassium, and calcium. Calcium helps to build strong bones, protein builds strong muscles, and potassium is essential for heart health.

The probiotics are the strongest health benefit that kefir has to offer.

- improve the healthy bacteria ratio in the gastrointestinal tract's environment
- treat or prevent diarrhea, especially following antibiotic treatment
- treat irritable bowel syndrome, or relieve symptoms
- reduce or prevent gastrointestinal infections, or increase recovery from them
- prevent and treat vaginal infections
- prevent and treat urinary tract infections

Kefir's health benefits even extend outside of its probiotics. One study found that mice who consumed kefir for seven days experienced anti-inflammatory and healing effects. Another study found that kefir may be beneficial in lowering cholesterol and stimulating the immune system.

Since kefir is safe to consume as a food, it may have fewer side effects than other probiotic supplements. It's also more nutritionally dense than other probiotic supplements. There is current research that focuses on many other benefits of a healthy gut biome with diverse bacteria.

Side Effects of Kefir

While kefir has plenty of great health benefits, it also has some side effects. These include constipation and abdominal cramping. These side effects are most common when you first start taking kefir.

Kefir is considered to be safe for children between 1 and 5 years of age, but ask their pediatrician first if you have concerns. Children under 1 year of age should not have cows' milk products, but breast milk is high in natural probiotics.

You should consult your doctor before drinking kefir if you have AIDS or other conditions that weaken the immune system. People with autoimmune diseases should also talk to their doctor before using kefir. While the bacteria in kefir is beneficial for people with strong immune systems, it may increase infections or exacerbations of conditions in those whose immune systems are out of balance.

Kefir is made with caseins, which some people eliminate from their diet. If you have eliminated caseins from your diet, it may be best to skip kefir and try another probiotic instead.

References

- Tortora, Gerard J.; Funke, Berdell R.; Case, Christine L. (2010). "5". Microbiology An Introduction (10 ed.). San Francisco, CA 94111, USA: Pearson Benjamin Cummings. p. 135. ISBN 978-0-321-58202-7
- Martini, A. (1992). "Biodiversity and conservation of yeasts". Biodiversity and Conservation. 1 (4): 324–333. doi:10.1007/BF00693768
- "Fermented fruits and vegetables. A global perspective". FAO Agricultural Services Bulletins - 134. Archived from the original on January 19, 2007. Retrieved 2007-01-28
- Ferry, J G (September 1992). "Methane from acetate". Journal of Bacteriology. 174 (17): 5489–5495. doi:10.1128/jb.174.17.5489-5495.1992. PMC 206491
- Achyut Yagnik; Suchitra Seth (24 August 2005). Shaping Of Modern Gujarat. Penguin Books Limited. p. 24. ISBN 978-81-8475-185-7
- Logan, BK; Distefano, S (1997). "Ethanol content of various foods and soft drinks and their potential for interference with a breath-alcohol test". Journal of analytical toxicology. 22 (3): 181–3. doi:10.1093/jat/22.3.181. PMID 9602932
- Ahmed Z, Wang Y, Ahmad A, Khan ST, Nisa M, Ahmad H, Afreen A (2013). "Kefir and health: a contemporary perspective". Crit Rev Food Sci Nutr. 53 (5): 422–34. doi:10.1080/10408398.2010.540360. PMID 23391011
- Hunziker, O F (January 1, 1923). "Utilization of Buttermilk in the form of Condensed and Dried Buttermilk" (PDF). Journal of Dairy Science. American Dairy Science Association. 6 (1): 1–12. doi:10.3168/jds.S0022-0302(23)94057-9. Retrieved 2010-10-26
- van Waarde, Aren; Thillart, G. Van den; Verhagen, Maria (1993). "Ethanol Formation and pH-Regulation in Fish". Surviving Hypoxia. pp. 157–170. ISBN 0-8493-4226-0
- Arslan, Seher (26 November 2014). "A review: chemical, microbiological and nutritional characteristics of kefir". CyTA – Journal of Food. 13 (3): 340–345. doi:10.1080/19476337.2014.981588

Chapter 4

Microbial Foodborne Diseases

Foodborne illness is a health condition that occurs due to the consumption of spoiled food. The major causes of food spoilage are pathogenic bacteria, parasites and viruses. This chapter discusses in detail about microbial foodborne diseases, infections and intoxications as well as the ways to prevent and treat them. All the diverse bacterial pathogens that cause food poisoning, such as Salmonella spp., Bacillus cereus, Campylobacter jejuni, Staphylococcus aureus, Escherichia coli, etc. have also been carefully analyzed in this chapter.

Foodborne Illness

Foodborne illnesses are infections or irritations of the gastrointestinal (GI) tract caused by food or beverages that contain harmful bacteria, parasites, viruses, or chemicals. The GI tract is a series of hollow organs joined in a long, twisting tube from the mouth to the anus. Common symptoms of foodborne illnesses include vomiting, diarrhea, abdominal pain, fever, and chills.

Most foodborne illnesses are acute, meaning they happen suddenly and last a short time, and most people recover on their own without treatment. Rarely, foodborne illnesses may lead to more serious complications. Each year, an estimated 48 million people in the United States experience a foodborne illness. Foodborne illnesses cause about 3,000 deaths in the United States annually.

Reasons for Foodborne Illnesses

The majority of foodborne illnesses are caused by harmful bacteria and viruses.[2] Some parasites and chemicals also cause foodborne illnesses.

Bacteria

Bacteria are tiny organisms that can cause infections of the GI tract. Not all bacteria are harmful to humans.

Some harmful bacteria may already be present in foods when they are purchased. Raw foods including meat, poultry, fish and shellfish, eggs, unpasteurized milk and dairy products, and fresh produce often contain bacteria that cause foodborne illnesses. Bacteria can contaminate food—making it harmful to eat—at any time during growth, harvesting or slaughter, processing, storage, and shipping.

Foods may also be contaminated with bacteria during food preparation in a restaurant or home kitchen. If food preparers do not thoroughly wash their hands, kitchen utensils, cutting boards, and other kitchen surfaces that come into contact with raw foods, cross-contamination—the spread of bacteria from contaminated food to uncontaminated food—may occur.

If hot food is not kept hot enough or cold food is not kept cold enough, bacteria may multiply. Bacteria multiply quickly when the temperature of food is between 40 and 140 degrees. Cold food should be kept below 40 degrees and hot food should be kept above 140 degrees. Bacteria multiply more slowly when food is refrigerated, and freezing food can further slow or even stop the spread of bacteria. However, bacteria in refrigerated or frozen foods become active again when food is brought to room temperature. Thoroughly cooking food kills bacteria.

Many types of bacteria cause foodborne illnesses. Examples include

- Salmonella, a bacterium found in many foods, including raw and undercooked meat, poultry, dairy products, and seafood. Salmonella may also be present on egg shells and inside eggs.
- Campylobacter jejuni (C. jejuni), found in raw or undercooked chicken and unpasteurized milk.
- Shigella, a bacterium spread from person to person. These bacteria are present in the stools of people who are infected. If people who are infected do not wash their hands thoroughly after using the bathroom, they can contaminate food that they handle or prepare. Water contaminated with infected stools can also contaminate produce in the field.
- Escherichia coli (E. coli), which includes several different strains, only a few of which cause illness in humans. E. coli O157:H7 is the strain that causes the most severe illness. Common sources of E. coli include raw or undercooked hamburger, unpasteurized fruit juices and milk, and fresh produce.
- Listeria monocytogenes (L. monocytogenes), which has been found in raw and undercooked meats, unpasteurized milk, soft cheeses, and ready-to-eat deli meats and hot dogs.
- Vibrio, a bacterium that may contaminate fish or shellfish.
- Clostridium botulinum (C. botulinum), a bacterium that may contaminate improperly canned foods and smoked and salted fish.

Viruses

Viruses are tiny capsules, much smaller than bacteria, that contain genetic material. Viruses cause infections that can lead to sickness. People can pass viruses to each other.

Viruses are present in the stool or vomit of people who are infected. People who are infected with a virus may contaminate food and drinks, especially if they do not wash their hands thoroughly after using the bathroom.

Common sources of foodborne viruses include

- food prepared by a person infected with a virus
- shellfish from contaminated water
- produce irrigated with contaminated water

Common foodborne viruses include

- norovirus, which causes inflammation of the stomach and intestines
- hepatitis A, which causes inflammation of the liver

Parasites

Parasites are tiny organisms that live inside another organism. In developed countries such as the United States, parasitic infections are relatively rare.

Cryptosporidium parvum and Giardia intestinalis are parasites that are spread through water contaminated with the stools of people or animals who are infected. Foods that come into contact with contaminated water during growth or preparation can become contaminated with these parasites. Food preparers who are infected with these parasites can also contaminate foods if they do not thoroughly wash their hands after using the bathroom and before handling food.

Trichinella spiralis is a type of roundworm parasite. People may be infected with this parasite by consuming raw or undercooked pork or wild game.

People Prone to Foodborne Illnesses

Anyone can get a foodborne illness. However, some people are more likely to develop foodborne illnesses than others, including

- infants and children
- pregnant women and their fetuses
- older adults
- people with weak immune systems

These groups also have a greater risk of developing severe symptoms or complications of foodborne illnesses.

Symptoms of Foodborne Illnesses

Symptoms of foodborne illnesses depend on the cause. Common symptoms of many foodborne illnesses include

- vomiting
- diarrhea or bloody diarrhea
- abdominal pain
- fever
- chills

Symptoms can range from mild to serious and can last from a few hours to several days.

C. botulinum and some chemicals affect the nervous system, causing symptoms such as

- headache
- tingling or numbness of the skin
- blurred vision
- weakness
- dizziness
- paralysis

Diagnosis of Foodborne Illnesses

To diagnose foodborne illnesses, health care providers ask about symptoms, foods and beverages recently consumed, and medical history. Health care providers will also perform a physical examination to look for signs of illness.

Diagnostic tests for foodborne illnesses may include a stool culture, in which a sample of stool is analyzed in a laboratory to check for signs of infections or diseases. A sample of vomit or a sample of the suspected food, if available, may also be tested. A health care provider may perform additional medical tests to rule out diseases and disorders that cause symptoms similar to the symptoms of foodborne illnesses.

If symptoms of foodborne illnesses are mild and last only a short time, diagnostic tests are usually not necessary.

Eating, Diet, and Nutrition

The following steps may help relieve the symptoms of foodborne illnesses and prevent dehydration in adults:

- drinking plenty of liquids such as fruit juices, sports drinks, caffeine-free soft drinks, and broths to replace fluids and electrolytes
- sipping small amounts of clear liquids or sucking on ice chips if vomiting is still a problem
- gradually reintroducing food, starting with bland, easy-to-digest foods such as rice, potatoes, toast or bread, cereal, lean meat, applesauce, and bananas
- avoiding fatty foods, sugary foods, dairy products, caffeine, and alcohol until recovery is complete

Infants and children present special concerns. Infants and children are likely to become dehydrated more quickly from diarrhea and vomiting because of their smaller body size. The following steps may help relieve symptoms and prevent dehydration in infants and children:

- giving oral rehydration solutions such as Pedialyte, Naturalyte, Infalyte, and CeraLyte to prevent dehydration
- giving food as soon as the child is hungry
- giving infants breast milk or fullstrength formula, as usual, along with oral rehydration solutions

Older adults and adults with weak immune systems should also drink oral rehydration solutions to prevent dehydration.

Traveler's Diarrhea

People who visit certain foreign countries are at risk for traveler's diarrhea, which is caused by eating food or drinking water contaminated with bacteria, viruses, or parasites. Traveler's diarrhea can be a problem for people traveling to developing countries in Africa, Asia, Latin America, and the Caribbean. Visitors to Canada, most European countries, Japan, Australia, and New Zealand do not face much risk for traveler's diarrhea.

To prevent traveler's diarrhea, people traveling from the United States to developing countries should avoid

- drinking tap water, using tap water to brush their teeth, or using ice made from tap water
- drinking unpasteurized milk or milk products
- eating raw fruits and vegetables, including lettuce and fruit salads, unless they peel the fruits or vegetables themselves
- eating raw or rare meat and fish

- eating meat or shellfish that is not hot when served
- eating food from street vendors

Travelers can drink bottled water, bottled soft drinks, and hot drinks such as coffee or tea.

People concerned about traveler's diarrhea should talk with a health care provider before traveling. The health care provider may recommend that travelers bring medication with them in case they develop diarrhea during their trip. Health care providers may advise some people—especially people with weakened immune systems—to take antibiotics before and during a trip to help prevent traveler's diarrhea. Early treatment with antibiotics can shorten a bout of traveler's diarrhea.

Types of Foodborne Illness

There are dozens of different bacteria, viruses, parasites and other pathogens which can contaminate raw or improperly prepared food thereby leading to human infection and the transmission of various foodborne diseases. However, the CDC estimates that about 90% of all foodborne illness in this country is caused by the following seven (7) pathogens: Norovirus, Salmonella, Clostridium perfrigens, Campylobacter, Listeria, E. coli 0157:H7 and Toxoplasma. While most people recover from food poisoning within a few days, every year nearly 130,000 Americans suffer a serious foodborne illnesses which requires hospitalization. In some cases, the foodborne disease proves fatal or leads to long term health consequences. The following are the most significant foodborne illnesses based upon frequency of occurrence and/or severity of potential adverse health consequences:

- Salmonellosis - a potentially life threatening infectious disease caused by various species of Salmonella bacteria. This infection can cause a debilitating condition referred to as Reactive Arthritis ("Reiter's Syndrome").
- Hemorrhagic Colitis - infectious disease primarily caused by the bacterium *Escherichia coli* O157:H7. E. coli infection can lead to a serious complication known as Hemolytic Uremic Syndrome (HUS).
- Listeriosis - infectious disease caused by the bacterium *Listeria monocytogenes*. This infection is particularly dangerous for pregnant women because it can cause stillbirth, premature labor and/or infection of the baby.
- Campylobacteriosis - infectious disease caused by the bacterium *Campylobacter jejuni*. This infection can lead to a serious disorder affecting the peripheral nervous system known as Guillain – Barre Syndrome.
- Shigellosis (bacillary dysentery) - infectious disease caused by various species of Shigella bacteria. This infection can also cause HUS or Reactive Arthritis.

- Perfringens Poisoning - enteric disease caused by the bacterium *Colostridium perfrigens.*
- Foodborne Botulism - life threatening disease caused by consumption of foods containing neurotoxins produced by the bacterium *Colostridium botulinum.*
- Norovirus Gastroenteritis - highly contagious disease caused by the Norwalk virus family. This viral infection can sometimes result in a chronic functional gastrointestinal disorder known as Post Infectious Irritable Bowel Syndrome.
- Acute Viral Hepatitis A - infectious disease of the liver caused by the Hepatitis A virus (HAV).
- Toxoplasmosis - infectious disease caused by a microscopic parasite called *Toxoplasma gondii.* In pregnant women, this infection can be transmitted and harmful to the fetus.

It is estimated that more than half of all foodborne illnesses are caused by some type of viral pathogen, while about one-third are caused by various species of bacteria. Parasites are responsible for a relatively small percentage (<5%) of foodborne illnesses in this country. Hepatitis A and Norovirus are highly contagious infectious diseases caused by viral pathogens. E. coli food poisoning, Salmonellis, Listeriosis, Campylobacteriosis and Shigellosis are communicable infectious diseases caused by different bacterial pathogens. Toxins produced by different species of bacteria in the Clostridium family cause Botulism and Perfringens poisoning which, unlike the other aforementioned illnesses, are not spread via person-to-person transmission from an infected individual to another previously non-infected individual. Toxoplasmosis is an infectious disease caused by a parasite. All of the pathogens which cause these illnesses – except Listeria, Hepatitis A and Toxoplasma – first attack the intestines causing acute gastrointestinal illness which may or may not lead to other illnesses or complications. Listeria, Hepatitis A and Toxoplasma infections generally cause various types of non-gastrointestinal illnesses. Listeria and Toxoplasma infections are unique in that they can be transmitted by an infected pregnant woman to her unborn child.

Some types of foodborne illness caused by bacteria can be treated with antibiotics with varying effectiveness. However, in certain types of bacterial food poisoning, antibiotics may be contraindicated. For example, the CDC recommends that E. coli infections not be treated with antibiotics because of a concern that they may increase the risk of Hemolytic Uremic Syndrome. Antibiotics are also not recommended for treatment of Perfringens poisoning. There is no "cure" for the viral infections, and they cannot be treated with antibiotics. However, acute viral Hepatitis A can be prevented with proper immunization. Unfortunately, Hepatitis A is the only one of these foodborne illnesses for which there currently is a vaccine.

The Most Dangerous Pathogens Causing Foodborne Illness

The most often occurring foodborne illnesses are Noroviruses, Salmonellosis, Perfringens poisoning and Campylobacteriosis, in that order. Fortunately, all of these foodborne illnesses, except Salmonellosis, have a very low mortality rate and incidence of serious complications; Salmonellosis results in significant morbidity. Indeed, Salmonella is the leading cause of death among all known foodborne pathogens. In addition to Salmonella, the foodborne pathogens of greatest concern include Toxoplasma gondii, Listeria monocytogenes, Norovirus, Campylobacter jejuni and E. coli (STEC) o157. According to the CDC, these six pathogens are to blame for more than 90% of all food poisoning deaths in the United States. Moreover, E. coli O157:H7 and Salmonella deserve particular attention because these pathogens are responsible for many of the most notorious food poisoning outbreaks which have taken place in the last two decades. Listeria and Toxoplasma pathogens are especially concerning for pregnant women because of their ability to infect both mother and fetus, and cause miscarriages, stillbirths or other complications during pregnancy.

The Potential Complications of Foodborne Illnesses and Persons at Highest Risk of Suffering Them

Fortunately, the majority of food poisoning episodes resolve within a week and without significant adverse health consequences. However, the risk of death, hospitalization and debilitating medical conditions is present with any foodborne illness, especially among certain groups of people considered most at risk. The categories of persons known to be at higher risk of contracting a foodborne illness and/or suffering death or serious medical complications therefrom include the following:

- Infants and young children
- Pregnant women (and their unborn fetus)
- Elderly adults over the age of 65
- Persons with a weakened or compromised immune system
- Cancer patients especially those undergoing chemotherapy or radiation treatments
- Persons who are HIV positive or suffering AIDS
- Persons with a certain chronic conditions such as diabetes, liver disease or kidney failure
- Persons who have sickle cell anemia or other disorders affecting red blood cells

- Transplant recipients especially while taking medications to reduce the risk of rejection
- Persons taking antibiotics, antihistamines or steroid medications

Food poisoning can be fatal, although the mortality rate in this country is very low. Death occurs in less than 1 out of every 10,000 cases of food poisoning. However, even with this low incident rate, the number of deaths is significant because the number of food poisoning episodes is so high (about 48 million cases are believed to occur annually). The U.S. food poisoning death toll is currently estimated by the CDC to be 3,000 per year, although some experts believe the annual number of deaths due to foodborne illness could be significantly higher. According to the USDA Economic Research Service, "foodborne illnesses account for about 1 of every 100 U.S. hospitalizations and for 1 of every 500 U.S. deaths." In cases of foodborne illness, the causes of death are varied but frequently include septicemia (which occurs after a bacterial infection has invaded the blood stream) and/or failure of one or more organs due to metastatic infection or some other disease process. For pregnant women, their unborn child is also at risk. Some foodborne illnesses (primarily Listeriosis and Toxoplasmosis) can cause spontaneous abortion and stillbirth of the fetus.

The potential health implications of a nonfatal foodborne illness are often under appreciated. According to the latest estimates released by the CDC in 2011, nearly 130,000 food poisoning victims require hospitalization in this country every year. Many of these patients face a wide array of potential complications with some having profound long term health consequences. Depending upon the type of foodborne illness involved, the list of known food poisoning complications includes diabetes, kidney dysfunction, urinary tract infections, renal induced hypertension, pernicious anaemia, hemorrhagic adrenalitis, partial paralysis, meningitis, encephalitis, brain damage, cognitive impairment, autoimmune disorders, glaucoma, vision impairment, endocarditis, pericarditis, lactose intolerance, gastroesophageal disease (GERD), dyspepsia, gastroparesis, and pseudomemdraneous colitis.

The Four Most Prevalent Food Poisoning Complications

Of particular note, the four most concerning and frequently cited medical conditions resulting from foodborne illnesses are as follows:

- Hemolytic Uremic Syndrome - a serious disease that occurs when an enteric bacterial infection leads to the production of toxins that enter the bloodstream and cause destruction of red blood cells. This can lead to further complications including hemolytic anemia, thrombocytopenia (blood platelet deficiency), uremia, pancreatitis, neurological dysfunction, seizures, acute respiratory distress syndrome (ARDS), renal damage and acute kidney failure. Children are at especially high risk for HUS. Most persons suffering HUS require hospitalization of signification duration. Most eventually recover but about one-third of patients suffer permanent sequellae, and the U.S. fatality rate is 3-5%. Most cases of HUS are associated with E. coli food poisoning, but this complication can also occur in cases of Shigelliosis.

- Guillain-Barre Syndrome (GBS) - a serious disorder of the peripheral nervous system caused by the body's autoimmune response to the presence of foreign pathogens. Among other things, GBS can lead to varying degrees of neuromuscular paralysis. Campylobacter is the primary cause of GBS.
- Reactive Arthritis (also called "Reiter's Syndrome") - a potentially debilitating disorder resulting from the body's autoimmune release of antigens to attack bacteria which have entered the bloodstream following a gastrointestinal infection. This process can cause joint inflammation and various other arthritic like conditions. This complication is known to occur in cases of food poisoning caused by Salmonella, Campylobacter, E. coli and Shigella.
- Post-Infectious Irritable Bowel Syndrome (IBS) - a potentially chronic and long-term functional gastrointestinal disorder characterized by various symptoms (which can occur daily or at irregular intervals) such as bloating, abdominal pain, diarrhea and constipation. This condition can be caused or exacerbated by severe gastrointestinal infections which occur after food poisoning. Foodborne illnesses caused by E. coli 0157:H7, Salmonella, Shigella, Campylobacter and Norovirus are known to be associated with IBS.

Foodborne Infection and Intoxication

Foodborne Infection

A foodborne infection is caused by the ingestion of food containing pathogenic microorganisms (i.e bacteria, virus or parasite) which must multiply with in the gastrointestinal tract, producing widespread inflammation. The most commonly implicated microorganisms include species of *Salmonella, Shigella, E. coli 0157:H7*, etc. These infections have longer incubation periods than those experienced with food intoxications, usually commencing from 6-24 hours or longer after ingestion. Symptoms may include fever, headache, nausea, vomiting, diarrhea, abdominal pain or distress, and prostration. The causative organism may be identified by laboratory examination of the vomitus, feces, or blood and the suspected food, when available.

Foods most commonly incriminated in outbreaks of foodborne infections are meat and seafood mixtures such as hash, hamburger, creamed meat pies, crab, lobster, chicken, and turkey salads, turkey, turkey stuffing or dressing, chicken, and ham. These foods have common characteristics in that they provide moisture, a good protein food supply and warmth. Given sufficient time, these factors promote an ideal environment for the growth and multiplication of microorganisms. It is important to remember that these organisms do not necessarily cause any alteration in the normal appearance, odor, or taste of the food.

Foodborne Intoxication

Certain bacteria under favorable growth conditions produce chemicals (toxins) in food which when ingested will cause food intoxication. Enterotoxins produced by *Staphylococcus aureus* are heat stable (i.e., not destroyed by normal cooking temperatures) and are the cause of the most common foodborne intoxication. The staphylococci multiply in the food where they produce their toxins before the food is consumed. It generally takes less than 8 hours for these organisms to elaborate enough toxins to cause symptoms. The disease is characterized by an abrupt onset (2 to 4 hours after ingestion) of symptoms of severe nausea, vomiting, diarrhea, and prostration with little or no fever.

Staphylococcal food intoxication usually follows ingestion of starchy food, especially potato salad, custard and pies. When the offending food is meat, pork (including ham and salami) and poultry products are usually the source. Ham may become contaminated with staphylococci during the practice of boning, slicing and holding without adequate refrigeration for several hours before serving. In addition, highly salted ham permits staphylococcal growth but inhibits many other bacteria. Other foods commonly involved are canned or potted meat or fish, pressed tongue, beef, cheese, other milk products, cream or custard filled pastries, potato salad, and pasta salads. The usual source of the pathogens, which cause this form of food intoxication, may be the nose, throat, boils, pimples, or infected cuts on the hands of food service personnel.

Prevention of Foodborne Illness

Following are the tips useful in preventing foodborne illnesses:

- Pick up your packaged and canned foods first.
- Don't buy food in cans that are bulging or dented or in jars that are cracked or have loose or bulging lids.
- Don't eat raw shellfish and use only pasteurized milk and cheese and pasteurized or otherwise treated ciders and juices if you have a health problem, especially one that may have impaired your immune system.
- Choose eggs that are refrigerated in the store. Before putting them in your cart, open the carton and make sure that the eggs are clean and none are cracked.
- Select frozen foods and perishables such as meat, poultry or fish last. Always

put these products in separate plastic bags so that drippings don't contaminate other foods in your shopping cart.

- Don't buy frozen seafood if the packages are open, torn or crushed on the edges. Avoid packages that are above the frost line in the store's freezer. If the package cover is transparent, look for signs of frost or ice crystals. This could mean that the fish has either been stored for a long time or thawed and refrozen.
- Check for cleanliness at the meat or fish counter and the salad bar. For instance, cooked shrimp lying on the same bed of ice as raw fish could become contaminated.
- When shopping for shellfish, buy from markets that get their supplies from state-approved sources; stay clear of vendors who sell shellfish from roadside stands or the back of a truck. And if you're planning to harvest your own shellfish, heed posted warnings about the safety of the water.
- Take an ice chest along to keep frozen and perishable foods cold if it will take more than an hour to get your groceries home.

Treatment of Foodborne Illness

Treatment for food poisoning typically depends on the source of the illness, if known, and the severity of your symptoms. For most people, the illness resolves without treatment within a few days, though some types of food poisoning may last longer.

Treatment of food poisoning may include:

- Replacement of lost fluids: Fluids and electrolytes — minerals such as sodium, potassium and calcium that maintain the balance of fluids in your body — lost to persistent diarrhea need to be replaced. Some children and adults with persistent diarrhea or vomiting may need hospitalization, where they can receive salts and fluids through a vein (intravenously), to prevent or treat dehydration.
- Antibiotics: Your doctor may prescribe antibiotics if you have certain kinds of bacterial food poisoning and your symptoms are severe. Food poisoning caused by listeria needs to be treated with intravenous antibiotics during hospitalization. The sooner treatment begins, the better. During pregnancy, prompt antibiotic treatment may help keep the infection from affecting the baby.

  Antibiotics will not help food poisoning caused by viruses. Antibiotics may actually worsen symptoms in certain kinds of viral or bacterial food poisoning. Talk to your doctor about your options.

Adults with diarrhea that isn't bloody and who have no fever may get relief from taking the medication loperamide (Imodium A-D) or bismuth subsalicylate (Pepto-Bismol). Ask your doctor about these options.

Lifestyle and Home Remedies

Food poisoning often improves without treatment within 48 hours. To help keep yourself more comfortable and prevent dehydration while you recover, try the following:

- Let your stomach settle: Stop eating and drinking for a few hours.
- Try sucking on ice chips or taking small sips of water: You might also try drinking clear soda, clear broth or noncaffeinated sports drinks, such as Gatorade. You're getting enough fluid when you're urinating normally and your urine is clear and not dark.
- Ease back into eating: Gradually begin to eat bland, low-fat, easy-to-digest foods, such as soda crackers, toast, gelatin, bananas and rice. Stop eating if your nausea returns.
- Avoid certain foods and substances until you're feeling better: These include dairy products, caffeine, alcohol, nicotine, and fatty or highly seasoned foods.
- Rest: The illness and dehydration can weaken and tire you.

Bacterial Pathogens

While most species of bacteria are harmless, and others beneficial for us, there are a number of disease-causing bacteria, which are called "pathogenic bacteria." Pathogenic bacteria can contribute to many worldwide diseases, including tuberculosis, cholera, anthrax, leprosy, the bubonic plague, pneumonia, and food-borne illnesses. The most common fatal bacterial infectious diseases are respiratory infections, with tuberculosis (caused by the bacteria *Mycobacterium tuberculosis*) killing approximately two million people each year, generally in sub-Saharan Africa. Bacterial infections can be caused by a wide range of bacteria which can lead to mild illness to life-threatening illnesses (like bacterial meningitis), which require immediate hospital interventions. Infections caused by bacteria are one of the leading causes of death in both the very young and the very old. Under normal circumstances, people are protected against bacterial infections by a healthy immune system.

Pathogenic bacteria use different strategies for infection: they produce toxins, the use parasites for spreading, or only by different adhesion strategies they are causing inflammation in the body.

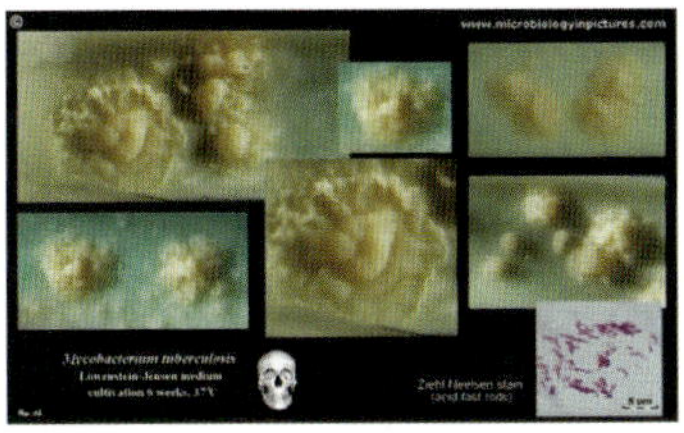

Mycobacterium tuberculosis the causative agent of tuberculosis

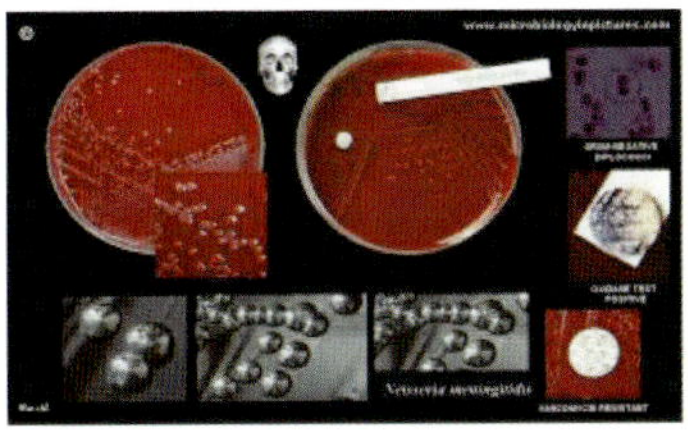

Neisseria meningitidis the causative agent of bacterial meningitis

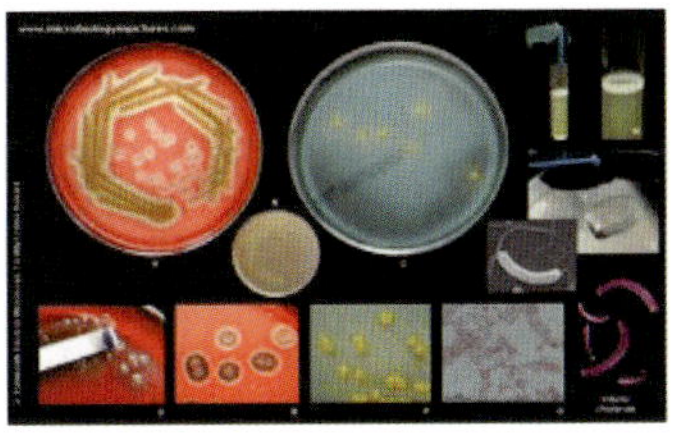

Vibrio cholerae the causative agent of cholera

Mechanisms of Damage

The symptoms of disease appear as pathogenic bacteria damage host tissues or interfere with their function. The bacteria can damage host cells directly. They can also cause damage indirectly by provoking an immune response that inadvertently damages host cells.

Direct

Once pathogens attach to host cells, they can cause direct damage as the pathogens use the host cell for nutrients and produce waste products. For example, *Streptococcus mutans*, a component of dental plaque, metabolizes dietary sugar and produces acid as a waste product. The acid decalcifies the tooth surface to cause dental caries. However, toxins produced by bacteria cause most of the direct damage to host cells.

Toxin Production

Endotoxins are the lipid portions of lipopolysaccharides that are part of the outer membrane of the cell wall of gram negative bacteria. Endotoxins are released when the bacteria lyses, which is why after antibiotic treatment, symptoms can worsen at first as the bacteria are killed and they release their endotoxins. Exotoxins are secreted into the surrounding medium or released when the bacteria die and the cell wall breaks apart.

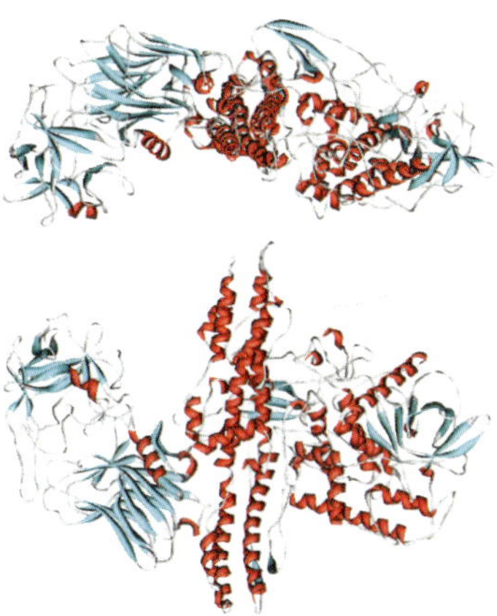

Protein structure of Botulinum toxin 3BTA

Indirect

An excessive or inappropriate immune response triggered by an infection may damage host cells.

Survival in Host

Nutrients

Iron is required for humans, as well as the growth of most bacteria. To obtain free iron, some pathogens secrete proteins called siderophores, which take the iron away from iron-transport proteins by binding to the iron even more tightly. Once the iron-siderophore complex is formed, it is taken up by siderophore receptors on the bacterial surface and then that iron is brought into the bacterium.

Identification

Typically identification is done by growing the organism in a wide range of cultures which can take up to 48 hours. The growth is then visually or genomically identified. The cultured organism is then subjected to various assays to observe reactions to help further identify species and strain.

Treatment

Bacterial infections may be treated with antibiotics, which are classified as bacteriocidal if they kill bacteria or bacteriostatic if they just prevent bacterial growth. There are many types of antibiotics and each class inhibits a process that is different in the pathogen from that found in the host. For example, the antibiotics chloramphenicol and tetracyclin inhibit the bacterial ribosome but not the structurally different eukaryotic ribosome, so they exhibit selective toxicity. Antibiotics are used both in treating human disease and in intensive farming to promote animal growth. Both uses may be contributing to the rapid development of antibiotic resistance in bacterial populations. Phage therapy can also be used to treat certain bacterial infections.

Prevention

Infections can be prevented by antiseptic measures such as sterilizing the skin prior to piercing it with the needle of a syringe and by proper care of indwelling catheters. Surgical and dental instruments are also sterilized to prevent infection by bacteria. Disinfectants such as bleach are used to kill bacteria or other pathogens on surfaces to prevent contamination and further reduce the risk of infection. Bacteria in food are killed by cooking to temperatures above 73 °C (163 °F).

Campylobacter Jejuni

Campylobacter jejuni: A bacterium that typically infects the bowels. Now the leading cause of bacterial food poisoning, Campylobacter jejuni is most often spread by contact with raw or undercooked poultry. A single drop of juice from a contaminated chicken is enough to make someone sick. Symptoms tend to start 2 to 5 days after exposure and typically last a week. They resemble the symptoms of viral gastroenteritis'diarrhea, fever, abdominal pain, cramping, nausea, and vomiting'but with campylobacter, fever is typical and the diarrhea is often bloody. Antibiotics can be helpful treatment. Most people recover completely. However, some suffer long-term consequences, such as arthritis or Guillain-Barr' syndrome. Both are thought to occur when a person's immune system is activated by the Campylobacter jejuni and misdirected to attack the person's own body.

Disease

Campylobacteriosis is an infectious disease caused by bacteria of the genus *Campylobacter*. In most people who become ill with campylobacteriosis, symptoms develop within two to five days of exposure to the organism and illness typically lasts seven days following onset. Infection with *C. jejuni* usually results in enteritis, which is characterised by abdominal pain, diarrhea, fever, and malaise. Diarrhea itself can vary in severity from loose to bloody stools. The disease is usually self-limiting. However, it does respond to antibiotics. Severe (accompanying fevers, blood in stools) or prolonged cases may require erythromycin, azithromycin, ciprofloxacin, or norfloxacin. Fluid replacement via Oral Rehydration Salts may be needed and intravenous fluid may be required for serious cases.

Pathogenesis

Studies on the pathogenesis of *C. jejuni* show that for this organism to cause disease, the susceptibility of the host and the relative virulence of the infecting strain are both important. Infection results from the ingestion of contaminated food or water, and the infective dose can be as low as 800 organisms. To initiate infection, the organism must penetrate the gastrointestinal mucus, which it does using its high motility and spiral shape. The bacteria must then adhere to the gut enterocytes and can then induce diarrhea by toxin release. *C. jejuni* releases several different toxins, mainly enterotoxin and cytotoxins, which vary from strain to strain and correlate with the severity of the enteritis. During infection, levels of all immunoglobulin classes rise. Of these, IgA is the most important because it can cross the gut wall. IgA immobilises organisms, causing them to aggregate and activate complement, and also gives short-term immunity against the infecting strain of organism. The bacteria colonize the small and large intestines, causing inflammatory diarrhea with fever. Stools contain leukocytes and blood. The role of toxins in pathogenesis is unclear. *C jejuni* antigens that cross-react with one or more neural structures may be responsible for triggering the Guillain–Barré syndrome.

Hypoacylated lipopolysaccharide (LPS) from *C. jejuni* induces moderate TLR4-mediated inflammatory response in macrophages and such LPS bioactivity may eventually result in the failure of local and systemic bacterial clearance in patients. At the same time, moderation of anti-bacterial responses may be advantageous for infected patients in clinical practice, since such an attenuated LPS may not be able to induce severe sepsis in susceptible individuals.

Sources

C. jejuni is commonly associated with poultry, and it naturally colonises the digestive tract of many bird species. All types of poultry and wild birds can become colonized with Campylobacter. One study found that 30% of European starlings in farm settings in Oxfordshire, United Kingdom, were carriers of *C. jejuni*. It is also common in cattle, and although it is normally a harmless commensal of the gastrointestinal tract in these animals, it can cause campylobacteriosis in calves. It has also been isolated from wombat and kangaroo feces, being a cause of bushwalkers' diarrhea. Contaminated drinking water and unpasteurized milk provide an efficient means for distribution. Contaminated food is a major source of isolated infections, with incorrectly prepared meat and poultry as the primary source of the bacteria. Moreover, surveys show that 20 to 100% of retail chickens are contaminated. This is not overly surprising, since many healthy chickens carry these bacteria in their intestinal tracts. Raw milk is also a source of infections. The bacteria are often carried by healthy cattle and by flies on farms. Unchlorinated water may also be a source of infections. However, properly cooking chicken, pasteurizing milk, and chlorinating drinking water kill the bacteria. *Campylobacter* is not, in contrast to *Salmonella*, transmitted vertically and therefore humans do not get infected by consuming eggs.

Possible Complications

Local complications of *Campylobacter* infections occur as a result of direct spread from the gastrointestinal tract and can include cholecystitis, pancreatitis, peritonitis, and massive gastrointestinal hemorrhage. Extraintestinal manifestations of *Campylobacter* infection are quite rare and may include meningitis, endocarditis, septic arthritis, osteomyelitis, and neonatal sepsis. Bacteremia is detected in <1% of patients with *Campylobacter* enteritis and is most likely to occur in patients who are immunocompromised or among the very young or very old. Transient bacteremia in immunocompetent hosts with *C. jejuni* enteritis may be more common, but not detected because most strains are rapidly cleared by the killing action of normal human serum and because blood cultures are not routinely performed for patients with acute gastrointestinal illness.

Serious systemic illness caused by *Campylobacter* infection rarely occurs, but can lead to sepsis and death. The case-fatality rate for *Campylobacter* infection is 0.05 per 1000 infections. For instance, one major possible complication that *C. jejuni* can cause is Guillain–Barré syndrome, which induces neuromuscular paralysis in a sizeable percentage of those who suffer from it. Over time, the paralysis is typically reversible to some extent; nonetheless, about 20% of patients with GBS are left disabled, and around 5% die. Another chronic condition that may be associated with *Campylobacter* infection is what was formerly known as Reiter's syndrome, a form of reactive arthritis. Reactive arthritis is a complication strongly associated with a particular genetic make-up. That is, persons who have the human lymphocyte antigen B27 (HLA-B27) are most susceptible. Most often, the symptoms of reactive arthritis will occur up to several weeks after infection.

Epidemiology

Frequency

- United States

An estimated 2 million cases of *Campylobacter* enteritis occur annually, accounting for 5–7% of cases of gastroenteritis. *Campylobacter* organisms have a large animal reservoir, with up to 100% of poultry, including chickens, turkeys, and waterfowl, having asymptomatic intestinal infections. The major reservoirs of *C. fetus* are cattle and sheep. Nonetheless, the incidence of *Campylobacter* infections has been declining. Changes in the incidence of culture-confirmed *Campylobacter* infections have been monitored by the Foodborne Diseases Active Surveillance Network (FoodNet) since 1996. In 2010, *Campylobacter* incidence showed a 27% decrease compared with 1996–1998. In 2010, the incidence was 13.6 cases per 100,000 population, and this did not change significantly compared with 2006–2008.

- International

C. jejuni infections are extremely common worldwide, although exact figures are not available. New Zealand reported the highest national campylobacteriosis rate, which peaked in May 2006 at 400 per 100,000 population.

- Sex

Campylobacter organisms are isolated more frequently from males than females. Homosexual men appear to be at increased risk for infection with atypical *Campylobacter* species such as *Helicobacter cinaedi* and *Helicobacter fennelliae.*

- Age

Campylobacter infections can occur in all age groups. Studies show a peak incidence in children younger than 1 year and in persons aged 15–29 years. The age-specific attack rate is highest in young children. In the United States, the highest incidence of *Campylobacter* infection in 2010 was in children younger than 5 years and was 24.4 cases per 100,000 population, However, the rate of fecal cultures positive for *Campylobacter* species is greatest in adults and older children.

Genome

The genome of *C. jejuni* strain NCTC11168 was published in 2000, revealing 1,641,481 base pairs (30.6% G+C) predicted to encode 1,654 proteins and 54 stable RNA species. The genome is unusual in that virtually no insertion sequences or phage-associated sequences and very few repeat sequences are found. One of the most striking findings in the genome was the presence of hypervariable sequences. These short homopolymeric runs of nucleotides were commonly found in genes encoding the biosynthesis or modification of surface structures, or in closely linked genes of unknown function. The apparently high rate of variation of these homopolymeric tracts may be important in the survival strategy of *C. jejuni*.

Initial transposon mutagenesis screens revealed 195 essential genes, although this number is likely to go up with additional analysis.

Natural Genetic Transformation

C. jejuni is naturally competent for genetic transformation. Natural genetic transformation is a sexual process involving DNA transfer from one bacterium to another through the intervening medium, and the integration of the donor sequence into the recipient genome by homologous recombination. *C. jejuni* freely takes up foreign DNA harboring genetic information responsible for antibiotic resistance. Antibiotic resistance genes are more frequently transferred in biofilms than between planktonic cells (single cells that float in liquid media).

Laboratory Characteristics

Characteristic	Result
Growth at 25 °C	–
Growth at 35–37 °C	+
Growth at 42 °C	+
Nitrate reduction	+
Catalase test	+
Oxidase test	+
Growth on MacConkey agar	+
Motility (wet mount)	+
Glucose utilization	–
Hippurate hydrolysis	+
Resistance to nalidixic acid	–
Resistance to cephalothin	+

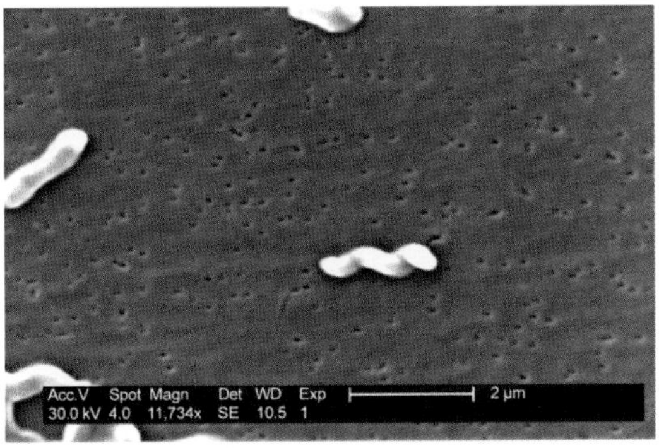

Scanning electron micrograph depicting a number of *Campylobacter jejuni* bacteria

Under light microscopy, *C. jejuni* has a characteristic "sea-gull" shape as a consequence of its helical form. *Campylobacter* is grown on specially selective "CAMP" agar plates at 42 °C, the normal avian body temperature, rather than at 37 °C, the temperature at which most other pathogenic bacteria are grown. Since the colonies are oxidase positive, they usually only grow in scanty amounts on the plates. Microaerophilic conditions are required for luxurious growth. A selective blood agar medium (Skirrow's medium) can be used. Greater selectivity can be gained with an infusion of a cocktail of antibiotics: vancomycin, polymixin-B, trimethoprim, and actidione ([Preston's agar]), and growth under microaerophilic conditions at 42 °C.

Symptoms

Some people who have a Campylobacter infection have mild symptoms, such as a few loose stools per day. These people probably would not seek medical attention, since the symptoms would resolve on their own.

When full-blown symptoms occur, they usually begin two to seven days after exposure to Campylobacter. At first, you may have a 12- to 48-hour period of fever, headache, muscle aches, and malaise (a general feeling of sickness). These early symptoms are followed by crampy abdominal pain and diarrhea, sometimes with nausea and vomiting. There may be up to 10 loose stools watery stools per day. You may notice some blood in the stool.

Diagnosis

Your doctor may suspect that you have some sort of food-related illness based on your history and symptoms. If your symptoms are very typical and not severe, the treatment for most food-related illnesses is the same. The only way to confirm that you definitely have been infected with *Campylobacter* is to collect a stool sample for examination in a laboratory (called a stool culture). However, a stool culture is often not needed.

Some patients have symptoms that are not typical. In those cases, the doctor would send the stool culture and also recommend additional tests such as CT scan.

Expected Duration

In adults who have normal immune defenses, *Campylobacter* infection is usually a self-limited illness that goes away on its own within seven to 10 days. With antibiotic treatment, symptoms often stop sooner, usually in about five days.

Prevention

Uncooked commercial chicken, raw veal and raw ground beef are frequent sources of Campylobacter. To help prevent infection:

Cook poultry and meat thoroughly

Wash your hands after handling raw poultry and meat

Wash kitchen countertops and utensils thoroughly after they have been used to prepare poultry and meat.

You also can reduce your risk by never drinking unpasteurized milk or untreated water.

If your pet becomes sick with diarrhea, wash your hands thoroughly after you have cared for it. Keep sick pets away from infants, elderly family members, and family members with weakened immune systems.

Bacillus Cereus

Bacillus cereus is a Gram-positive, rod-shaped, aerobic, facultatively anaerobic, motile, beta hemolytic bacterium commonly found in soil and food. Some strains are harmful to humans and cause foodborne illness, while other strains can be beneficial as probiotics for animals. It is the cause of "fried rice syndrome", as the bacteria are classically contracted from fried rice dishes that have been sitting at room temperature for hours. *B. cereus* bacteria are facultative anaerobes, and like other members of the genus *Bacillus*, can produce protective endospores. Its virulence factors include cereolysin and phospholipase C.

The *Bacillus cereus* group comprises seven closely related species: *B. cereus sensu stricto* (referred to herein as *B. cereus*), *B. anthracis*, *B. thuringiensis*, *B. mycoides*, *B. pseudomycoides*, *B. weihenstephanensis*, and *B. cytotoxicus*.

Ecology

B. cereus competes with other microorganisms such as *Salmonella* and *Campylobacter* in the gut, so its presence reduces the numbers of those microorganisms. In food animals such as chickens, rabbits and pigs, some harmless strains of *B. cereus* are used as a probiotic feed additive to reduce *Salmonella* in the intestines and cecum. This improves the animals' growth, as well as food safety for humans who eat them. They parasitize codling moth larvae.

B. cereus and other members of *Bacillus* are not easily killed by alcohol; in fact, they have been known to colonize distilled liquors and alcohol-soaked swabs and pads in numbers sufficient to cause infection.

Some strains of *B. cereus* produce cereins, bacteriocins active against different *B. cereus* strains or other Gram-positive bacteria.

Reproduction

At 30 °C (86 °F), a population of *B. cereus* can double in as little as 20 minutes or as long as 3 hours, depending on the food product.

Food	Minutes to double, 30 °C (86 °F)	Hours to multiply by 1,000,000
Milk	20–36	6.6 - 12
Cooked rice	26–31	8.6 - 10.3
Infant formula	56	18.6

Pathogenesis

B. cereus is responsible for a minority of foodborne illnesses (2–5%), causing severe nausea, vomiting, and diarrhea. *Bacillus* foodborne illnesses occur due to survival of the bacterial endospores when food is improperly cooked. Cooking temperatures less than or equal to 100 °C (212 °F) allow some *B. cereus* spores to survive. This problem is compounded when food is then improperly refrigerated, allowing the endospores to germinate. Cooked foods not meant for either immediate consumption or rapid cooling and refrigeration should be kept at temperatures below 10 °C (50 °F) or above 50 °C (122 °F). Germination and growth generally occur between 10 °C and 50 °C, though some strains are psychrotrophic. Bacterial growth results in production of enterotoxins, one of which is highly resistant to heat and acids (pH levels between 2 and 11); ingestion leads to two types of illness, diarrheal and emetic (vomiting) syndrome.

- The diarrheal type is associated with a wide range of foods, has an 8.0- to 16-hour incubation time, and is associated with diarrhea and gastrointestinal pain. Also known as the 'long-incubation' form of *B. cereus* food poisoning, it might be difficult to differentiate from poisoning caused by *Clostridium perfringens*. Enterotoxin can be inactivated after heating at 56 °C (133 °F) for 5 minutes however it is unclear whether its presence in food causes the symptom since it degrades in stomach enzymes; its subsequent production by surviving *B. cereus* spores within the small intestine may be the cause of illness.
- The 'emetic' form is commonly caused by rice cooked for a time and temperature insufficient to kill any spores present, then improperly refrigerated. It can produce a toxin, cereulide, which is not inactivated by later reheating. This form leads to

nausea and vomiting one to five hours after consumption. It can be difficult to distinguish from other short-term bacterial foodborne intoxications such as by *Staphylococcus aureus*. Emetic toxin can withstand 121 °C (250 °F) for 90 minutes.

The diarrhetic syndromes observed in patients are thought to stem from the three toxins: hemolysin BL (Hbl), nonhemolytic enterotoxin (Nhe) and cytotoxin K (CytK). The *nhe/hbl/cytK* genes are located on the chromosome of the bacteria. Transcription of these genes is controlled by *PlcR*. These genes occur in the taxonomically related *B. thuringiensis* and *B. anthracis*, as well. These enterotoxins are all produced in the small intestine of the host, thus thwarting digestion by host endogenous enzymes. The Hbl and Nhe toxins are pore-forming toxins closely related to ClyA of *E. coli*. The proteins exhibit a conformation known as "beta-barrel" that can insert into cellular membranes due to a hydrophobic exterior, thus creating pores with hydrophilic interiors. The effect is loss of cellular membrane potential and eventually cell death. CytK is a pore-forming protein more related to other hemolysins.

The timing of the toxin production was previously thought to be possibly responsible for the two different courses of disease, but in fact the emetic syndrome is caused by a toxin, cereulide, found only in emetic strains and is not part of the "standard toolbox" of *B. cereus*. Cereulide is a cyclic polypeptide containing three repeats of four amino acids: D-oxy-Leu—D-Ala—L-oxy-Val—L-Val (similar to valinomycin produced by *Streptomyces griseus*) produced by nonribosomal peptide synthesis. Cereulide is believed to bind to 5-hydroxytryptamine 3 (5-HT3) serotonin receptors, activating them and leading to increased afferent vagus nerve stimulation. It was shown independently by two research groups to be encoded on multiple plasmids: pCERE01 or pBCE4810. Plasmid pBCE4810 shares homology with the *Bacillus anthracis* virulence plasmid pXO1, which encodes the anthrax toxin. Periodontal isolates of *B. cereus* also possess distinct pXO1-like plasmids. Like most of cyclic peptides containing nonproteogenic amino acids, cereulid is resistant to heat, proteolysis, and acid conditions.

B. cereus is also known to cause difficult-to-eradicate chronic skin infections, though less aggressive than necrotizing fasciitis. *B. cereus* can also cause keratitis.

Diagnosis

In case of foodborne illness, the diagnosis of *B. cereus* can be confirmed by the isolation of more than 10^5 *B. cereus* organisms per gram from epidemiologically implicated food, but such testing is often not done because the illness is relatively harmless and usually self-limiting.

Identification Through Testing

Below is a list of differential techniques and results that can help to identify *Bacillus cereus* from other bacteria and *Bacillus* species.

- Anaerobic Growth: Positive
- Voges Proskauer test: Positive
- Acid produced from
  - D-glucose: Positive
  - L-arabinose: Negative
  - D-xylose: Negative
  - D-mannitol: Negative
- Starch hydrolysis: Positive
- Nitrate reduction: Positive
- Degradation of tyrosine: Positive
- Growth at
  - >50 °C: Negative
- Utilization of citrate: Positive

Prognosis

Most emetic patients recover within six to 24 hours, but in some cases, the toxin can be fatal. In 2014, 23 neonates receiving total parenteral nutrition contaminated with *B. cereus* developed septicaemia, with three of the infants later dying as a result of infection.

Bacteriophage

Bacteria of the *B. cereus* group are infected by bacteriophage belonging to the family Tectiviridae. This family includes tail-less phages that have a lipid membrane/vesicle beneath the icosahedral protein shell and that are formed of approximately equal amounts of virus-encoded proteins and lipids derived from the host cell plasma membrane. Upon infection, the lipid membrane becomes a tail-like structure used in genome delivery. The genome is composed of ~15-kb linear double-stranded DNA (dsDNA) with long inverted terminal repeat sequences (100 bp). GIL01, Bam35, GIL16, AP50, and Wip1 are examples of temperate tectiviruses infecting the *B. cereus* group.

Yersinia Enterocolitica

Yersinia enterocolitica is a bacterium that causes an infectious disease called yersiniosis. Yersinia enterocolitica is a member of the Yersinia family of bacteria.

Yersiniosis is most often acquired by eating contaminated food, especially raw or undercooked pork products. The preparation of raw pork intestines (chitterlings) may be particularly risky. Infants can be infected if their caretakers handle raw chitterlings and then do not adequately clean their hands before handling the infant or the infant's toys, bottles, or pacifiers. Drinking contaminated unpasteurized milk or untreated water can also transmit the infection. Occasionally Y. enterocolitica infection occurs after contact with infected animals. On rare occasions, it can be transmitted as a result of the bacterium passing from the stools or soiled fingers of one person to the mouth of another person. This may happen when basic hygiene and handwashing habits are inadequate. Rarely, the organism is transmitted through contaminated blood during a transfusion.

Common symptoms of yersiniosis in children (who most often contract the disease) are fever, abdominal pain, and diarrhea, which is often bloody. Symptoms typically develop 4 to 7 days after exposure and may last 1 to 3 weeks or longer. In older children and adults, right-sided abdominal pain and fever may be the predominant symptoms and may be confused with appendicitis. In a small proportion of cases, complications such as skin rash, joint pains, or spread of bacteria to the bloodstream can occur.

Yersiniosis is generally diagnosed by detecting the organism in the stool. The organism can also be recovered from other sites, including the throat, lymph nodes, joint fluid, urine, bile, and blood.

Uncomplicated cases of diarrhea due to Y. enterocolitica usually resolve on their own without antibiotic treatment. However, in more severe or complicated infections, antibiotics such as aminoglycosides, doxycycline, trimethoprim-sulfamethoxazole, or fluoroquinolones may be useful.

Most infections with Y. enterocolitica are uncomplicated and resolve completely. Occasionally, some persons develop joint pain, most commonly in the knees, ankles or wrists. These joint pains usually develop about 1 month after the initial episode of diarrhea and generally resolve after 1 to 6 months. A skin rash, called "erythema nodosum," may also appear on the legs and trunk; this is more common in women. In most cases, erythema nodosum resolves spontaneously within a month.

To prevent the infection:

1. Avoid eating raw or undercooked pork.
2. Consume only pasteurized milk or milk products.
3. Wash hands with soap and water before eating and preparing food, after contact with animals, and after handling raw meat.
4. After handling raw chitterlings, clean hands and fingernails scrupulously with soap and water before touching infants or their toys, bottles, or pacifiers.

Someone other than the foodhandler should care for children while chitterlings are being prepared.

5. Prevent cross-contamination in the kitchen:
   - Use separate cutting boards for meat and other foods.
   - Carefully clean all cutting boards, counter-tops, and utensils with soap and hot water after preparing raw meat.
6. Dispose of animal feces in a sanitary manner.

Treatment

Yersiniosis is usually self-limiting and does not require treatment. For sepsis or severe focal infections, especially if associated with immunosuppression, the recommended regimen includes doxycycline in combination with an aminoglycoside. Other antibiotics active against *Y. enterocolitica* include trimethoprim-sulfamethoxasole, fluoroquinolones, ceftriaxone, and chloramphenicol. *Y. enterocolitica* is usually resistant to penicillin G, ampicillin, and cefalotin due to beta-lactamase production.

Staphylococcus Aureus

Staphylococcus aureus is a gram positive bacterium that usually appears under the microscope as spherical (coccus) organisms appearing in pairs, short chains, or bunched, grape-like clusters.

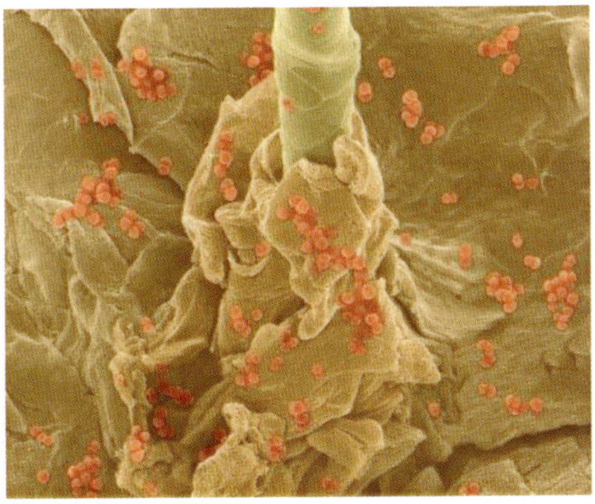

Dennis Kunkel Microscopy

Diseases caused by Staphylococcus Aureus

Staphylococcus aureus has been implicated as a causative agent in acute food poisoning episodes, toxic shock syndrome, impetigo, scalded skin syndrome, cellulitis, folliculitis and furuncles.

It is also a common cause of systemic infections such as infective endocarditis, osteomyelitis, epiglottitis, and sinus infections amongst others. S. aureus is also responsible for many infective and systemic infections in the health care setting (nosocomial infections).

People Susceptible to Infections with Staphylococcus Aureus

All people are susceptible to infections with Staphylococcus aureus, the disease however, may manifest with varying intensities.

Epidemiology of Staphylococcus Aureus

Staphylococcus aureus can cause illness by preformed toxin production as well as by infecting both local tissues and the systemic circulation. Disease transmission can occur in gastrointestinal. Staphylococcus aureus causes acute episodes of food poisoning via preformed enterotoxins. Food items likely to be infected by staphylococcal food poisoning include meat and meat products; poultry and egg products; salads such as egg, tuna, chicken, potato, and macaroni; bakery products such as cream-filled pastries, cream pies, and chocolate éclairs; sandwich fillings; and milk and dairy products.

Salmonella spp.

Salmonella spp. are a group of bacteria which reside in the intestinal tract of human beings and warm blooded animals and are capable of causing disease. They are the second most common cause of bacterial foodborne illness in Ireland (*Campylobacter* spp. is the most frequent cause). They are facultative anaerobic Gramnegative rods.

Salmonella spp. are members of the *Enterobacteriaceae* group. The genus *Salmonella* contains 2 species:

- *Salmonella enterica*
- *Salmonella bongori*

Salmonella enterica is an important agent of foodborne illness. This species is sub-classified into 6 subspecies of which *S. enterica* subspecies enterica is the most important for human health.

The genus *Salmonella* can be subdivided into more than 2,400 serotypes. *Salmonella enterica* subsp. *Enterica* serotype Typhimurium (*S.* Typhimurium) and *Salmonella enterica* subsp. *enterica* serotype Enteritidis (*S.* Enteritidis) are the most frequently isolated serotypes in humans in Ireland. Serotypes are further subdivided by their resistance to bacteriophages (phage types or lystotypes), antibiotics or heavy metals; their biochemical characteristics (biovars or biotypes) or their sensitivity to or production of bacteriocins.

Growth and Survival Characteristics

Table: Factors affecting the growth of *Salmonella* spp.

Conditions	Minimum	Optimum	Maximum
Temperature (°C)	5.2	35 – 43	46.2
pH	3.8	7 – 7.5	9.5
Water activity (a_W)	0.94	0.99	>0.99

Salmonella spp. are not particularly heat resistant and most serotypes are killed by normal cooking conditions, i.e. cooking to a core temperature of 75°C instantaneously or an equivalent time temperature combination, e.g. 70°C for 2 minutes. However, a few highly heat resistant serotypes have been reported, e.g. *S.* Senftenberg 775W and *S.* Irumu. Heat resistance is influenced by water activity (aw), nature of the solutes and pH of the suspending medium. Greater heat resistance is observed for cells in sucrose compared with NaCl at the same aw values. Heat resistance also increases with increasing concentration of milk solids.

Heat resistance of *S.* Typhimurium in milk chocolate at various temperatures is presented in the following table:

	D-Value (Min) For S. Typhimurium In Milk Chocolate
90°C	75
80°C	222
70°C	816

Heat resistance is usually expressed by the D-value (i.e. the time in minutes at a given temperature to achieve a 90% reduction in the number of viable cells). ICMSF 1996 provides an extensive list of D-Values for *Salmonella* spp. under different conditions and food types.

Sources of *Salmonella* spp. and Routes of Transmission

Salmonella spp. reside in the intestinal tract of humans and warm blooded animals. They are shed in the faeces. Poor hygiene practices can result in the spread of this pathogen to the hands of humans while on the farm. Feet, hair and skin of animals can become contaminated as they walk, sit or lie in faecally contaminated ground or litter.

The spread of *Salmonella* spp. to carcasses may occur in the abattoir environment during slaughter. Furthermore, spread of *Salmonella* spp. to agricultural commodities may occur through faecal contamination of streams, rivers and coastal waters and the use of improperly treated sewage for agricultural purposes. In the food processing environment, spread of this pathogen can occur through cross contamination from raw food or infected food handlers.

The human infection caused by *Salmonella* spp. is referred to as salmonellosis. Although salmonellosis can arise from contact with infected animals, consumption of contaminated food is the most usual cause.

Salmonella spp. and Food

Many foods have been identified as vehicles for the transmission of this pathogen to humans, particularly foods of animal origin but also foods of non-animal origin which may be subject to faecal contamination. Those of particular importance include eggs, pig meat, poultry meat, milk, chocolate, fruit and vegetables. It is important to note that although *Salmonella* spp. do not form spores, they can survive for long periods in foods.

As pathogens

Salmonella species are facultative intracellular pathogens. A facultative organism uses oxygen to make ATP; when it is not available it "exercises its option"—the literal meaning of the term—and makes ATP via fermentation, or by substituting one or more of four less efficient electron acceptors as oxygen at the end of the electron transport chain: sulfate, nitrate, sulfur, or fumarate.

Most infections are due to ingestion of food contaminated by animal feces, or by human feces, such as by a food-service worker at a commercial eatery. *Salmonella* serotypes can be divided into two main groups—typhoidal and nontyphoidal. Nontyphoidal serotypes are more common, and usually cause self-limiting gastrointestinal disease. They can infect a range of animals, and are zoonotic, meaning they can be transferred between humans and other animals. Typhoidal serotypes include *Salmonella* Typhi and *Salmonella* Paratyphi A, which are adapted to humans and do not occur in other animals.

Nontyphoidal Salmonella

Infection with nontyphoidal serotypes of *Salmonella* generally results in food poisoning. Infection usually occurs when a person ingests foods that contain a high concentration of the bacteria. Infants and young children are much more susceptible to infection, easily achieved by ingesting a small number of bacteria. In infants, infection through inhalation of bacteria-laden dust is possible.

The organisms enter through the digestive tract and must be ingested in large numbers to cause disease in healthy adults. An infection can only begin after living salmonellae (not merely *Salmonella*-produced toxins) reach the gastrointestinal tract. Some of the microorganisms are killed in the stomach, while the surviving ones enter the small intestine and multiply in tissues. Gastric acidity is responsible for the destruction of the majority of ingested bacteria, but *Salmonella* has evolved a degree of tolerance to acidic environments that allows a subset of ingested bacteria to survive. Bacterial colonies may also become trapped in mucus produced in the esophagus. By the end of the incubation period, the nearby host cells are poisoned by endotoxins released from the dead salmonellae. The local response to the endotoxins is enteritis and gastrointestinal disorder.

About 2,000 serotypes of nontyphoidal *Salmonella* are known, which may be responsible for as many as 1.4 million illnesses in the United States each year. People who are at risk for severe illness include infants, elderly, organ-transplant recipients, and the immunocompromised.

Invasive Nontyphoidal Salmonella Disease

While in developed countries, nontyphoidal serotypes present mostly as gastrointestinal disease, in sub-Saharan Africa, these serotypes can create a major problem in bloodstream infections, and are the most commonly isolated bacteria from the blood of those presenting with fever. Bloodstream infections caused by nontyphoidal salmonellae in Africa were reported in 2012 to have a case fatality rate of 20–25%. Most cases of invasive nontyphoidal salmonella infection (iNTS) are caused by *S. typhimurium* or *S. enteritidis*. A new form of *Salmonella typhimurium* (ST313) emerged in the southeast of the African continent 75 years ago, followed by a second wave which came out of central Africa 18 years later. This second wave of iNTS possibly originated in the Congo Basin, and early in the event picked up a gene that made it resistant to the antibiotic chloramphenicol. This created the need to use expensive antimicrobial drugs in areas of Africa that were very poor, making treatment difficult. The increased prevalence of iNTS in sub-Saharan Africa compared to other regions is thought to be due to the large proportion of the African population with some degree of immune suppression or impairment due to the burden of HIV, malaria, and malnutrition, especially in children. The genetic makeup of iNTS is evolving into a more typhoid-like bacterium, able to efficiently spread around the human body. Symptoms are reported to be diverse, including fever, hepatosplenomegaly, and respiratory symptoms, often with an absence of gastrointestinal symptoms.

Typhoidal Salmonella

Typhoid fever is caused by *Salmonella* serotypes which are strictly adapted to humans or higher primates—these include *Salmonella* Typhi, Paratyphi A, Paratyphi B, and Paratyphi C. In the systemic form of the disease, salmonellae pass through the lymphatic system of the intestine into the blood of the patients (typhoid form) and are carried to various organs (liver, spleen, kidneys) to form secondary foci (septic form). Endotoxins first act on the vascular and nervous apparatus, resulting in increased permeability and decreased tone of the vessels, upset of thermal regulation, and vomiting and diarrhoea. In severe forms of the disease, enough liquid and electrolytes are lost to upset the water-salt metabolism, decrease the circulating blood volume and arterial pressure, and cause hypovolemic shock. Septic shock may also develop. Shock of mixed character (with signs of both hypovolemic and septic shock) is more common in severe salmonellosis. Oliguria and azotemia may develop in severe cases as a result of renal involvement due to hypoxia and toxemia.

Global Monitoring

In Germany, food poisoning infections must be reported. From 1990 to 2016, the number of officially recorded cases decreased from about 200,000 to about 13,000 cases. In the United States, about 1,200,000 cases of *Salmonella* infections are reported each year. A World Health Organization study estimated that 21,650,974 cases of typhoid fever occurred in 2000, 216,510 of which resulted in death, along with 5,412,744 cases of paratyphoid fever.

Molecular Mechanisms of Infection

Mechanisms of infection differ between typhoidal and nontyphoidal serotypes, owing to their different targets in the body and the different symptoms that they cause. Both groups must enter by crossing the barrier created by the intestinal cell wall, but once they have passed this barrier, they use different strategies to cause infection.

Nontyphoidal serotypes preferentially enter M cells on the intestinal wall by bacterial-mediated endocytosis, a process associated with intestinal inflammation and diarrhoea. They are also able to disrupt tight junctions between the cells of the intestinal wall, impairing the cells' ability to stop the flow of ions, water, and immune cells into and out of the intestine. The combination of the inflammation caused by bacterial-mediated endocytosis and the disruption of tight junctions is thought to contribute significantly to the induction of diarrhoea.

Salmonellae are also able to breach the intestinal barrier via phagocytosis and trafficking by CD18-positive immune cells, which may be a mechanism key to typhoidal *Salmonella* infection. This is thought to be a more stealthy way of passing the intestinal barrier, and may, therefore, contribute to the fact that lower numbers of typhoidal *Salmonella* are required for infection than nontyphoidal *Salmonella*. *Salmonella* cells are able to enter macrophages via macropinocytosis. Typhoidal serotypes can use this to achieve dissemination throughout the body via the mononuclear phagocyte system, a network of connective tissue that contains immune cells, and surrounds tissue associated with the immune system throughout the body.

Much of the success of *Salmonella* in causing infection is attributed to two type III secretion systems which function at different times during an infection. One is required for the invasion of nonphagocytic cells, colonization of the intestine, and induction of intestinal inflammatory responses and diarrhea. The other is important for survival in macrophages and establishment of systemic disease. These systems contain many genes which must work co-operatively to achieve infection.

The AvrA toxin injected by the SPI1 type III secretion system of *S.* Typhimurium works to inhibit the innate immune system by virtue of its serine/threonine acetyltransferase activity, and requires binding to eukaryotic target cell phytic acid (IP6). This leaves the host more susceptible to infection.

Salmonellosis is known to be able to cause back pain or spondylosis. It can manifest as five clinical patterns: gastrointestinal tract infection, enteric fever, bacteremia, local infection, and the chronic reservoir state. The initial symptoms are nonspecific fever, weakness, and myalgia among others. In the bacteremia state, it can spread to any parts of the body and this induces localized infection or it forms abscesses. The forms of localized *Salmonella* infections are arthritis, urinary tract infection, infection of the central nervous system, bone infection, soft tissue infection, etc. Infection may remain as the latent form for a long time, and when the function of reticular endothelial cells is deteriorated, it may become activated and consequently, it may secondarily induce spreading infection in the bone several months or several years after acute salmonellosis.

Resistance to Oxidative Burst

A hallmark of *Salmonella* pathogenesis is the ability of the bacterium to survive and proliferate within phagocytes. Phagocytes produce DNA damaging agents such as nitric oxide and oxygen radicals as a defense against pathogens. Thus, *Salmonella* species must face attack by molecules that challenge genome integrity. Buchmeier *et al.* showed that mutants of *S. enterica* lacking RecA or RecBC protein function are highly sensitive to oxidative compounds synthesized by macrophages, and furthermore these findings indicate that successful systemic infection by *S. enterica* requires RecA and RecBC mediated recombinational repair of DNA damage.

Host Adaptation

S. enterica, through some of its serotypes such as Typhimurium and Enteriditis, shows signs of the ability to infect several different mammalian host species, while other serotypes such as Typhi seem to be restricted to only a few hosts. Some of the ways that *Salmonella* serotypes have adapted to their hosts include loss of genetic material and mutation. In more complex mammalian species, immune systems, which include pathogen specific immune responses, target serovars of *Salmonella* through binding of antibodies to structures like flagella. Through the loss of the genetic material that codes for a flagellum to form, *Salmonella* can evade a host's immune system. *mgtC* leader RNA from bacteria virulence gene (mgtCBR operon) decreases flagellin production during infection by directly base pairing with mRNAs of the *fljB* gene encoding flagellin and promotes degradation. In the study by Kisela *et al.*, more pathogenic serovars of *S. enterica* were found to have certain adhesins in common that have developed out of convergent evolution. This means that, as these strains of *Salmonella* have been exposed to similar conditions such as immune systems, similar structures evolved separately to negate these similar, more advanced defenses in hosts. There are still many questions about the way that *Salmonella* has evolved into so many different types but it has been suggested that *Salmonella* evolved through several phases. As Baumler *et al.* have suggested, Salmonella most likely evolved through horizontal gene transfer, formation of new serovars due to additional pathogenicity islands and through an approximation of its ancestry. So, *Salmonella* could have evolved into its many different serotypes through gaining genetic information from different pathogenic bacteria. The presence of several pathogenicity islands in the genome of different serotypes has lent credence to this theory.

Genetics

In addition to its importance as a pathogen, *S. enterica* serovar Typhimurium has been instrumental in the development of genetic tools that led to an understanding of fundamental bacterial physiology. These developments were enabled by the discovery of the first generalized transducing phage, P22, in Typhimurium that allowed quick and easy genetic exchange that allowed fine structure genetic analysis. The large number of mutants led to a revision of genetic nomenclature for bacteria. Many of the uses of transposons as genetic tools, including transposon delivery, mutagenesis, construction of chromosome rearrangements, were also developed in Typhimurium. These genetic tools also led to a simple test for carcinogens, the Ames Test.

Vibrio spp.

Vibrio, (genus Vibrio), any of a group of comma-shaped bacteria in the family Vibrionaceae. Vibrios are aquatic microorganisms, some species of which cause serious diseases in humans and other animals.

Vibrios are microbiologically characterized as gram-negative, highly motile, facultative anaerobes (not requiring oxygen), with one to three whiplike flagella at one end. Their cells are curved rods 0.5 μm (micrometre; 1 μm = 10^{-6} metre) across and 1.5 to 3.0 μm long, single or strung together in S-shapes or spirals.

Three species of vibrio are of significance to humans: V. cholerae is the cause of cholera, and V. parahaemolyticus and V. vulnificus both act as agents of acute enteritis, or bacterial diarrhea. V. anguillarum is found in diseased eels and other fishes.

Ways People Get Vibrosis

Most people become infected by eating raw or undercooked shellfish, particularly oysters. Certain Vibrio species can also cause a skin infection when an open wound is exposed to brackish or salt water. Brackish water is a mixture of fresh and sea water. It is often found where rivers meet the sea.

People Prone to Vibriosis

People with compromised immune systems, especially those with chronic liver disease, are more likely to get vibriosis. Eating raw seafood, particularly oysters, and exposing open wounds to brackish or salt water can increase a person's chance for getting vibriosis. About 80% of infections occur between May and October when water temperatures are warmer.

How Common is Vibriosis?

It is estimated that vibriosis causes 80,000 illnesses each year in the United States. About 52,000 of these illnesses are estimated to be the result of eating contaminated food.

The most commonly reported species, Vibrio parahaemolyticus, is estimated to cause 45,000 illnesses each year in the United States.

Is Vibriosis a Serious Disease?

Most people with a mild case of vibriosis recover after about 3 days with no lasting effects. However, people with a Vibrio vulnificus infection can get seriously ill and need intensive care or limb amputation. About 1 in 4 people with this type of infection die, sometimes within a day or two of becoming ill.

Prevention of Vibriosis

To reduce your chance of getting vibriosis, don't eat raw or undercooked shellfish, such as oysters. If you have a wound (including cuts and scrapes), avoid contact with brackish or salt water or cover the wound with a waterproof bandage if there's a possibility it could come into contact with brackish or salt water, raw seafood, or raw seafood juices.

Clostridium spp.

It is a microbe that lives right here on earth. This microbe is a bacterium that is soil born and can be found in the intestines of animals.

Clostridium sporogenes, or C. sporogenes for short, is a Gram-positive bacteria. What is that you might ask? A Gram-positive bacterium is one that stains violet in color during a Gram stain test. This means that it has a very thick outer cell wall made up of peptidoglycan, which is a sugar structure that creates a protective layer around the bacteria to keep is safe from attack.

Clostridia are gram-positive spore-forming bacilli that cause up to 10% of anaerobic infections and occur at diverse clinical sites. Of the more than 200 recognized species of Clostridium, <20 are associated with invasive and toxigenic human diseases. Clostridial species vary in oxygen tolerance, nutritional needs, motility, and optimal temperature for growth. They can cause distinct clostridial histotoxic syndromes due to specific clostridial toxins (e.g., gas gangrene, food poisoning) and other, nonsyndromic clostridial infections (e.g., bloodstream infection, BSI). Clostridia also are recovered from various sites as part of polymicrobial infection, including intra-abdominal, biliary tract, female genital tract, pleuropulmonary, central nervous system, and skin and soft-tissue infections.

Clostridium contains around 100 species that include common free-living bacteria, as well as important pathogens. The main species responsible for disease in humans are:

- *Clostridium botulinum* can produce botulinum toxin in food or wounds and can cause botulism. This same toxin is known as Botox and is used in cosmetic surgery to paralyze facial muscles to reduce the signs of aging; it also has numerous other therapeutic uses.
- *Clostridium difficile* can flourish when other gut flora bacteria are killed during antibiotic therapy, leading to superinfection and potentially fatal pseudomembranous colitis (a severe necrotizing disease of the large intestine).
- *Clostridium perfringens* causes a wide range of symptoms, from food poisoning to cellulitis, fasciitis, and gas gangrene.
- *Clostridium tetani* causes tetanus.
- *Clostridium sordellii* can cause a fatal infection in exceptionally rare cases after medical abortions.

Bacillus and *Clostridium* are often described as gram-variable, because they show an increasing number of gram-negative cells as the culture ages.

Clostridium and *Bacillus* are both in the division Firmicutes, but they are in different classes, orders, and families. Microbiologists distinguish *Clostridium* from *Bacillus* by the following features:

- *Clostridium* grows in anaerobic conditions; *Bacillus* grows in aerobic conditions.
- *Clostridium* forms bottle-shaped endospores; *Bacillus* forms oblong endospores.
- *Clostridium* does not form the enzyme catalase; *Bacillus* secretes catalase to destroy toxic byproducts of oxygen metabolism.

Clostridium and *Desulfotomaculum* are both in the class Clostridia and order Clostridiales, and they both produce bottle-shaped endospores, but they are in different families. *Clostridium* can be distinguished from *Desulfotomaculum* on the basis of the nutrients each genus uses (the latter requires sulfur).

Glycolysis and fermentation of pyruvic acid by Clostridia yield the end products butyric acid, butanol, acetone, isopropanol, and carbon dioxide.

The Schaeffer-Fulton stain (0.5% malachite green in water) can be used to distinguish endospores of *Bacillus* and *Clostridium* from other microorganisms. There is a commercially available polymerase chain reaction (PCR) test kit (Bactotype) for the detection of *C. perfringens* and other pathogenic bacteria.

Treatment

In general, the treatment of clostridial infection is high-dose penicillin G, to which the organism has remained susceptible. *Clostridium welchii* and *Clostridium tetani* respond to sulfonamides. Clostridia are also susceptible to tetracyclines, carbapenems (imipenem), metronidazole, vancomycin, and chloramphenicol.

The vegetative cells of clostridia are heat-labile and are killed by short heating at temperatures above 72–75 °C. The thermal destruction of *Clostridium* spores requires higher temperatures (above 121.1 °C, for example in an autoclave) and longer cooking times (20 min, with a few exceptional cases of > 50 min recorded in the literature). *Clostridia* and *Bacilli* are quite radiation-resistant, requiring doses of about 30 kGy, which is a serious obstacle to the development of shelf-stable irradiated foods for general use in the retail market. The addition of lysozyme, nitrate, nitrite and propionic acid salts inhibits clostridia in various foods.

Fructooligosaccharides (fructans) such as inulin, occurring in relatively large amounts in a number of foods such as chicory, garlic, onion, leek, artichoke, and asparagus, have a prebiotic or bifidogenic effect, selectively promoting the growth and metabolism of beneficial bacteria in the colon, such as bifidobacteria and lactobacilli, while inhibiting harmful ones, such as clostridia, fusobacteria, and bacteroides.

Clostridium Botulinum

Botulism is a rare but potentially life-threatening bacterial illness. *Clostridium Botulinum* bacteria grows on food and produces toxins that, when ingested, cause paralysis. Botulism poisoning is extremely rare, but so dangerous that each case is considered a public health emergency. Studies have shown that there is a 35 to 65 percent chance of death for patients who are not treated immediately and effectively with botulism antitoxin. Infant botulism is the most common form of botulism.

Most of the botulism cases reported each year come from foods that are not canned properly at home. Botulism from commercially canned food is rare, but commercial canned chili products were identified as the source of a botulism outbreak in 2007.

Symptoms of Botulism

Botulism neurotoxins prevent neurotransmitters from functioning properly. This means that they inhibit motor control. As botulism progresses, the patient experiences paralysis from top to bottom, starting with the eyes and face and moving to the throat, chest, and extremities. When paralysis reaches the chest, death from inability to breathe results unless the patient is ventilated. Symptoms of botulism generally appear 12 to 72 hours after eating contaminated food. With treatment, illness lasts from 1 to 10 days. Full recovery from botulism poisoning can take weeks to months. Some people never fully recover.

In general, symptoms of botulism poisoning include the following:

- Nausea
- Vomiting
- Fatigue
- Dizziness
- Double vision
- Dry skin, mouth and throat
- Drooping eyelids
- Difficulty swallowing
- Slurred speech
- Muscle Weakness
- Body Aches
- Paralysis
- Lack of fever

Infant botulism takes on a different form. Symptoms in an infant include lethargy, poor appetite, constipation, drooling, drooping eyelids, a weak cry, and paralysis.

Long-Term Effects of Botulism

The majority of botulism patients never fully recover their pre-illness health. After three months to a year of recovery, persisting side-effects are most likely permanent. These long-term effects most often include fatigue, weakness, dizziness, dry mouth, and difficulty performing strenuous tasks. Patients also report a generally less happy and peaceful psychological state than before their illness.

Botulism Diagnosis

If a patient displays symptoms of botulism, a doctor will most likely take a blood, stool, or gastric secretion sample. The most common test for botulism is injecting the patient's blood into a mouse to see whether the mouse displays signs of botulism, since other testing methods take up to a week.

Sometimes botulism can be difficult to diagnose, since symptoms can be mild, or confused with those of Guillan-Barre Syndrome.

Treatment of Botulism

If found early, botulism can be treated with an antitoxin that blocks circulation of the toxin in the bloodstream. This prevents the patient's case from worsening, but recovery still takes several weeks.

Clostridium Perfringens

Clostridium perfringens (formerly known as *C. welchii*, or *Bacillus welchii*) is a Gram-positive, rod-shaped, anaerobic, spore-forming pathogenic bacterium of the genus *Clostridium*. *C. perfringens* is everpresent in nature and can be found as a normal component of decaying vegetation, marine sediment, the intestinal tract of humans and other vertebrates, insects, and soil. It has the shortest reported generation time of any organism at 6.3 minutes in thioglycolate medium.

C. perfringens is one of the most common causes of food poisoning in the United States, alongside norovirus, *Salmonella*, *Campylobacter*, and *Staphylococcus aureus*. However, it can sometimes be ingested and cause no harm.

Infections due to *C. perfringens* show evidence of tissue necrosis, bacteremia, emphysematous cholecystitis, and gas gangrene, which is also known as clostridial myonecrosis. The toxin involved in gas gangrene is known as α-toxin, which inserts into the plasma membrane of cells, producing gaps in the membrane that disrupt normal cellular function. *C. perfringens* can participate in polymicrobial anaerobic infections. *Clostridium perfringens* is commonly encountered in infections as a component of the normal flora. In this case, its role in disease is minor.

The action of *C. perfringens* on dead bodies is known to mortuary workers as tissue gas and can be halted only by embalming.

Genome

C. perfringens has a stable G+C content of approximately 27–28% and average genome size of 3.5 Mb. Genomes of 56 *C. perfringens* strains have since been made available on NCBI genomes database for scientific research community. Genomic research has revealed surprisingly high diversity in *C. perfringens* pangenome, with only 12.6% core genes, identified as the most divergent Gram-positive bacteria reported. Nevertheless, 16S rRNA regions in between *C. perfringens* strains are found to be highly conserved (sequence identity >99.1%).

Motility

Although they lack flagella, *C. perfringens* bacteria are able to glide across surfaces because their bodies are lined with filaments from end-to-end. The hypermotile variants such as SM101, are often found arising on the edges of colonies on agar plates. Video microscopy of their gliding movement suggest that they form long, thin filaments that allow them to move rapidly like bacteria with flagella. Genome sequencing was used to identify the cause(s) of the hypermotile phenotype and their direct derivatives. In comparing them, strains SM124 and SM127, hypermotile derivatives of strains SM101 and SM102, respectively, contained 10 and 6 nucleotide polymorphisms (SNPs) relative to their parent strains. Mutations in cell division genes is the common feature of the hypermotile strains.

Food Poisoning

In the United Kingdom and United States, *C. perfringens* bacteria are the third most common cause of foodborne illness, with poorly prepared meat and poultry, or food properly prepared but left to stand too long, the main culprits in harboring the bacterium. The clostridium perfringens enterotoxin (CPE) mediating the disease is heat-labile (inactivated at 74 °C (165 °F)). It can be detected in contaminated food (if not heated properly), and feces. Incubation time is between six and 24 (commonly 10–12) hours after ingestion of contaminated food.

Since *C. perfringens* forms spores that can withstand cooking temperatures, if cooked food is let stand for long enough, germination can ensue and infective bacterial colonies develop. Symptoms typically include abdominal cramping, diarrhea, vomiting, and fever. The whole course usually resolves within 24 hours, but can last up to two weeks in older or infirm hosts.

Perfringens poisoning can also lead to another disease known as enteritis necroticans (also known as pigbel); this is caused by *C. perfringens* type c. However this infection is often fatal. Large numbers of *C. perfringens* grow in the intestines, and secrete exotoxin. This exotoxin causes necrosis of the intestines, varying levels of hemorrhaging and perforation of the intestine. Inflammation usually occurs in sections of the jejunum, midsection of the small intestine. This disease eventually leads to septic shock and death. This particular disease is rare in the United States; typically it occurs in populations with a higher risk. Risk factors for enteritis necroticans include protein-deficient diet, unhygienic food preparation, sporadic feasts of meat (after long periods of a protein-deficient diet), diets containing large amounts of trypsin inhibitors (sweet potatoes), areas prone to infection of the parasite Ascaris (produces a trypsin inhibitor). This disease is contracted in populations living in New Guinea, parts of Africa, Central America, South America and Asia.

Many cases of *C. perfringens* food poisoning likely remain subclinical, as antibodies to the toxin are common among the population. This has led to the conclusion that most of the population has experienced food poisoning due to *C. perfringens*.

Despite its potential dangers, *C. perfringens* is used as the leavening agent in salt rising bread. The baking process is thought to reduce the bacterial contamination, precluding negative effects.

Infection

Clostridium perfringens is the most common bacterial agent for gas gangrene. Some symptoms include blisters, tachycardia, swelling, and jaundice.

Research published in 2014 suggested that a strain of *C. perfringens* might be implicated in multiple sclerosis (MS). Tests in mice found that a toxin made by a rare strain of *C. perfringens* caused MS-like damage in the brain, and earlier work had identified this strain of *C. perfringens* in a human with MS. MS patients were found to be ten times more immune-reactive to the epsilon toxin than healthy people.

Diagnosis

C. perfringens can be diagnosed by Nagler's reaction where the suspect organism is cultured on an egg yolk media plate. One side of the plate contains anti-alpha-toxin, while the other side does not. A streak of suspect organism is placed through both sides. An area of turbidity will form around the side that does not have the anti-alpha-toxin, indicating uninhibited lecithinase activity. In addition, laboratories can diagnose the bacteria by determining the number of bacteria in the feces. Within the 48 hours from when the disease began, if the individual has more than 106 spores of the bacteria per gram of stool, then the illness is diagnosed as *C. perfringens* food poisoning. Other tests/reactions: Catalase: Negative, Spot indole: Positive, Lecithinase: Positive, Lipase: Negative, Litmus Milk: Stormy Fermentation, Reverse CAMP plate: Positive, Gas Liquid Chromatography products: (Acetic, Butyric and Lactic Acids).

Typically, the symptoms of *C. perfingens* poisoning are used to diagnose it. However, diagnosis can be made using a stool culture test, in which the feces is tested for toxins produced by the bacteria.

Prevention

The growth of *C. perfringens* spores can be prevented by most importantly cooking food, especially beef and poultry, thoroughly to the recommended temperatures. Leftover food should be refrigerated to a temperature below 40 °F (4 °C) within two hours of preparation. Large pots of food like soup or stew with meats should be divided in small quantities and covered for refrigeration. Leftovers should be reheated to at least 165 °F (74 °C) before serving. A rule of thumb is that if the food tastes, smells, or looks different from what it is supposed to then the food should be avoided. Even if it looks okay, a food that has been out for a long time can also be dangerous to eat.

Food Poisoning Incidents

On May 7, 2010, "42 residents and 12 staff members at a Louisiana state psychiatric hospital were affected experienced vomiting, abdominal cramps, and diarrhea". Three patients died within 24 hours. The outbreak was linked to chicken which was cooked a day prior to being served and was not cooled down according to hospital guidelines. The outbreak affected 31% of the residents of the hospital and 69% of the staff who ate the chicken. It is unknown how many of the affected residents ate the chicken.

In May 2011, a man died after allegedly eating food contaminated with the bacteria on a transatlantic American Airlines flight. The man's wife and daughter are suing American and LSG Sky Chefs, the German company that prepared the inflight food.

In December 2012, a 46-year-old woman died two days after eating a Christmas Day meal at a pub in Hornchurch, Essex, England. She was among about 30 people to fall ill after eating the meal. Samples taken from the victims contained *C. perfringens*. The hotel manager and the cook were jailed for offences arising from the incident.

In December 2014, 87-year-old Bessie Scott died three days after eating a church pot-luck supper in Nackawic, New Brunswick, Canada. Over 30 other people reported signs of gastrointestinal illness, diarrhea and abdominal pain. The province's acting chief medical officer says *Clostridium perfringens* is the bacteria that most likely caused the woman's death.

In October 2016, 66 year-old Alex Zdravich died four days after eating an enchilada, burrito, and taco at Agave Azul in West Lafayette, Indiana. Three others that dined the same day reported signs of foodborne illness, which were consistent with the symptoms and rapid onset of *Clostridium perfringens*. They later tested positive for the presence of the bacteria.

In November 2016, food contaminated with *Clostridium perfringens* caused three individuals to die, and another 22 to be sickened, after a Thanksgiving luncheon hosted by a church in Antioch, California.

In January 2017, a mother and her son sued a restaurant in Rochester, New York as they and 260 other people were sickened after eating foods contaminated with

Clostridium perfringens. "Officials from the Monroe County Department of Public Health closed down the Golden Ponds after more than a fourth of its Thanksgiving Day guests became ill. An inspection revealed a walk-in refrigerator with food spills and mold, a damaged gasket preventing the door from closing, and mildew growing inside."

Epidemiology

C. perfringens is the leading cause of food poisoning in the United States and Canada. Contaminated meats in stews, soups and gravies are usually responsible for outbreaks and causes about 250,000 cases of foodborne illnesses in the United States every year. Deaths due to the disease are rare and mostly occurs in elderly and people who are predisposed to the disease. From 1998–2010, 289 confirmed outbreaks of *C. perfringens* illness were reported with 15,208 illnesses, 82 hospitalizations, and 8 deaths. Outbreaks can cause regularly and can cause substantial morbidity but it can be prevented by avoiding the contamination of meats that were coming from the slaughterhouses and farms.

Shigella spp.

Shigella is the bacterium that causes the disease shigellosis, also known as bacillary dysentery. *Shigella* is one of the most easily transmitted bacterial diarrheas, since it can occur after fewer than 100 bacteria are ingested. While reported cases of *Shigella* range between 14,000 and 20,000 annually, with the majority of these cases occurring between July and October. *Shigella Sonnei* is the most common type of *Shigella*. It accounts for over two-thirds of cases of shigellosis in the United States.

Shigella bacteria are generally transmitted through a fecal-oral route. Foods that come into contact with human or animal waste can transmit *Shigella*. Thus, handling toddlers' diapers, eating vegetables from a field contaminated with sewage, or drinking pool water are all activities that can lead to shigellosis.

Symptoms of Shigella Food Poisoning

Symptoms of *Shigella* poisoning most commonly develop one to three days after exposure to *Shigella* bacteria, and usually go away within five to seven days. It is also possible to get *Shigella* but experience no symptoms, and still be contagious to others, a condition known as being asymptomatic.

Common *Shigella* Food Poisoning Symptoms

- Diarrhea: Diarrhea ranges from mild to severe. It is bloody in 25 to 50 percent of cases and usually contains mucus
- Fever

- Stomach cramps
- Rectal spasms

Complications from Shigella

Complications from shigellosis can include severe dehydration, seizures in small children, rectal bleeding, and invasion of the blood stream by the bacteria. Young children and the elderly are at the highest risk of death. The following is a list of specific complications caused by *Shigella*.

Proctitis and Rectal Prolapse: The bacteria that causes shingellosis can also cause inflammation of the lining of the rectum or rectal prolapse.

Reactive Arthritis: Approximately 3 percent of patients with *Shigella* infection, most often those with *Shigella flexneri*, develop Reactive Arthritis. It occurs when the immune system attempts to combat *Shigella* but instead attacks the body. Symptoms of Reactive Arthritis include inflammation of the joints, eyes, or reproductive or urinary organs. On average, symptoms appear 18 days after infection.

Toxic Megacolon: In this rare complication, the colon is paralyzed and unable to pass bowel movements or gas. Symptoms of Toxic Megacolon include abdominal pain and swelling, fever, weakness, and disorientation. If this complication goes untreated and the colon ruptures, the patient's condition can be life-threatening.

Hemolytic Uremic Syndrome, or HUS: *Shigella* rarely results in HUS, which is more commonly a complication of Shiga toxin-producing E. coli infections. HUS can lead to kidney failure.

Diagnosis of Shigella

A *Shigella* infection is diagnosed through laboratory testing of a stool sample.

Shigella Food Poisoning Treatment

A *Shigella* infection usually goes away on its own in five to seven days, although bowel movements may continue to be abnormal for up to a month following infection. Antibiotics, however, can shorten the course of the illness. A doctor can prescribe antibiotics after testing a stool sample for the presence of *Shigella* bacteria.

Some strains of shigellosis are resistant to antibiotics, meaning that antibiotics might not always be an effective treatment. Antidiarrheal medication should be avoided, as it can actually make the illness worse.

Listeria Monocytogenes

Listeria monocytogenes is the species of pathogenic bacteria that causes the infection listeriosis. It is a facultative anaerobic bacterium, capable of surviving in the presence or absence of oxygen. It can grow and reproduce inside the host's cells and is one of the most virulent foodborne pathogens, with 20 to 30% of food borne listeriosis infections in high-risk individuals may be fatal. Responsible for an estimated 1,600 illnesses and 260 deaths in the United States (U.S.) annually, listeriosis ranks third in total number of deaths among food borne bacterial pathogens, with fatality rates exceeding even *Salmonella* and *Clostridium botulinum*. In the European Union listeriosis follows an upward trend that began in 2008, causing 2,161 confirmed cases and 210 reported deaths in 2014, 16% more than in 2013. Listeriosis mortality rates in the US are also higher in the EU than for other food-borne pathogens.

Listeria monocytogenes is a Gram-positive bacterium, in the division Firmicutes, named after Joseph Lister. Its ability to grow at temperatures as low as 0 °C permits multiplication at typical refrigeration temperatures, greatly increasing its ability to evade control in human foodstuffs. Motile via flagella at 30 °C and below, but usually not at 37 °C, *L. monocytogenes* can instead move within eukaryotic cells by explosive polymerization of actin filaments (known as comet tails or actin rockets).

Studies suggest up to 10% of human gastrointestinal tracts may be colonized by *Listeria monocytogenes*.

Nevertheless, clinical diseases due to *Listeria monocytogenes* are more frequently recognized by veterinarians, especially as meningoencephalitis in ruminants.

Due to its frequent pathogenicity, causing meningitis in newborns (acquired transvaginally), pregnant mothers are often advised not to eat soft cheeses such as Brie, Camembert, feta, and queso blanco fresco, which may be contaminated with and permit growth of *Listeria monocytogenes*. It is the third-most-common cause of meningitis in newborns. *Listeria monocytogenes* can infect the brain, spinal cord membranes and/or the bloodstream of the host through the ingestion of contaminated food such as unpasteurized dairy or raw foods.

Classification

Listeria monocytogenes is a Gram-positive, non spore-forming, motile, facultatively anaerobic, rod-shaped bacterium. It is catalase-positive and oxidase-negative, and expresses a beta hemolysin, which causes destruction of red blood cells. This bacterium exhibits characteristic tumbling motility when viewed with light microscopy. Although *L. monocytogenes* is actively motile by means of peritrichous flagella at room temperature (20–25 °C), the organism does not synthesize flagella at body temperatures (37 °C).

The genus *Listeria* belongs to the class Bacilli and the order Bacillales, which also includes *Bacillus* and *Staphylococcus*. *Listeria* currently contains 10 species: *L. fleischmannii, L. grayi, L. innocua, L. ivanovii, L. marthii, L. monocytogenes,*

L. rocourtiae, *L. seeligeri*, *L. weihenstephanensis* and *L. welshimeri*. *L. denitrificans*, previously thought to be part of the genus *Listeria*, was reclassified into the new genus *Jonesia*. Both *L. ivanovii* and *L. monocytogenes* are pathogenic in mice, but only *L. monocytogenes* is consistently associated with human illness. The 13 serotypes of *L. monocytogenes* can cause disease, but more than 90% of human isolates belong to only three serotypes: 1/2a, 1/2b, and 4b. *L. monocytogenes* serotype 4b strains are responsible for 33 to 5% of sporadic human cases worldwide and for all major foodborne outbreaks in Europe and North America since the 1980s.

Pathogenesis

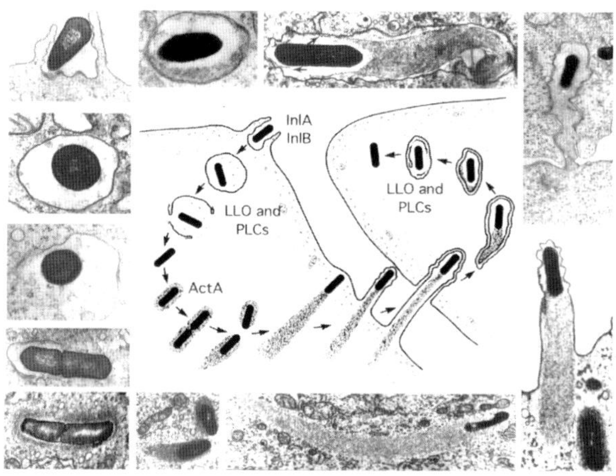

Stages in the intracellular lifecycle of *L. monocytogenes*. (Center) Cartoon depicting entry, escape from a vacuole, actin nucleation, actin-based motility, and cell-to-cell spread. (Outside) Representative electron micrographs from which the image was derived.

Invasive infection by *L. monocytogenes* causes the disease listeriosis. When the infection is not invasive, any illness as a consequence of infection is termed febrile gastroenteritis. The manifestations of listeriosis include septicemia, meningitis (or meningoencephalitis), encephalitis, corneal ulcer, pneumonia, and intrauterine or cervical infections in pregnant women, which may result in spontaneous abortion (second to third trimester) or stillbirth. Surviving neonates of fetomaternal listeriosis may suffer granulomatosis infantiseptica — pyogenic granulomas distributed over the whole body — and may suffer from physical retardation. Influenza-like symptoms, including persistent fever, usually precede the onset of the aforementioned disorders. Gastrointestinal symptoms, such as nausea, vomiting, and diarrhea, may precede more serious forms of listeriosis or may be the only symptoms expressed. Gastrointestinal symptoms were epidemiologically associated with use of antacids or cimetidine. The onset time to serious forms of listeriosis is unknown, but may range from a few days to three weeks. The onset time to gastrointestinal symptoms is unknown but probably exceeds 12 hours. An early study suggested that *L. monocytogenes* is unique among Gram-positive bacteria in that it might possess lipopolysaccharide, which serves as an endotoxin. Later, it was found to not be a true endotoxin. *Listeria* cell walls consistently contain lipoteichoic acids, in which a glycolipid moiety, such as a galactosyl-glucosyl-diglyceride, is covalently linked to the terminal phosphomonoester of the teichoic acid. This lipid region anchors the polymer chain to the cytoplasmic membrane. These lipoteichoic acids resemble the lipopolysaccharides of Gram-negative bacteria in both structure and function, being the only amphipathic polymers at the cell surface.

L. monocytogenes has D-galactose residues on its surface that can attach to D-galactose receptors on the host cell walls. These host cells are generally M cells and Peyer's patches of the intestinal mucosa. Once attached to this cells, *L. monocytogenes* can translocate past the intestinal membrane and into the body.

The infective dose of *L. monocytogenes* varies with the strain and with the susceptibility of the victim. From cases contracted through raw or supposedly pasteurized milk, one may safely assume that, in susceptible persons, fewer than 1,000 total organisms may cause disease. *L. monocytogenes* may invade the gastrointestinal epithelium. Once the bacterium enters the host's monocytes, macrophages, or polymorphonuclear leukocytes, it becomes bloodborne (septicemic) and can grow. Its presence intracellularly in phagocytic cells also permits access to the brain and probably transplacental migration to the fetus in pregnant women. The pathogenesis of *L. monocytogenes* centers on its ability to survive and multiply in phagocytic host cells. It seems that *Listeria* originally evolved to invade membranes of the intestines, as an intracellular infection, and developed a chemical mechanism to do so. This involves a bacterial protein "internalin"(InlA/InlB) which attaches to a protein on the intestinal cell membrane "cadherin" and allows the bacteria to invade the cells through a zipper mechanism. These adhesion molecules are also to be found in two other unusually tough barriers in humans — the blood-brain barrier and the fetal–placental barrier, and this may explain the apparent affinity that *Listeria* has for causing meningitis and affecting babies *in utero*. Once inside the cell, Listeria rapidly acidifies the lumen of the vacuole formed around it during cell entry to activate listeriolysin O, a cholesterol-dependent cytolysin capable of disrupting the vacuolar membrane. This frees the pathogen and gives it access to the cytosol of the cell, where it continues its pathogenesis. Motility in the intracellular space is provided by actin assembly-inducing protein (ActA) which allows the bacteria to use the host cell's actin polymerization machinery to polymerize the cytoskeleton to give a "boost" to the bacterial cell so it can move in the cell. The same ActA mechanism also allows the bacteria to travel from cell to cell.

Regulation of Pathogenesis

L. monocytogenes can act as a saprophyte or a pathogen, depending on its environment. When this bacterium is present within a host organism, quorum sensing causes the up-regulation of several virulence genes. Depending on the location of the bacterium within the host organism, different activators up-regulate the virulence genes. SigB, an alternative sigma factor, up-regulates *Vir* genes in the intestines, whereas PrfA up-regulates gene expression when the bacterium is present in blood. Little is known about how this bacterium switches between acting as a saprophyte and a pathogen; however, several noncoding RNAs are thought to be required to induce this change.

Pathogenicity of Lineages

L. monocytogenes has three distinct lineages, with differing evolutionary histories and pathogenic potentials. Lineage I strains contain the majority of human clinical isolates and all human epidemic clones, but are underrepresented in animal clinical isolates. Lineage II strains are overrepresented in animal cases and underrepresented in human clinical cases, and are more prevalent in environmental and food samples. Lineage III isolates are very rare, but significantly more common in animal than human isolates.

Detection

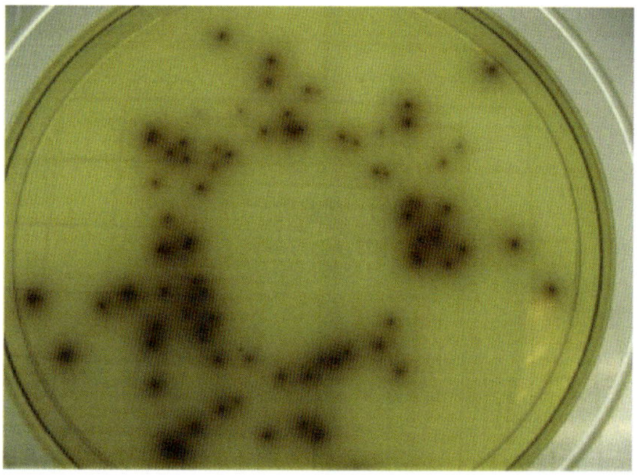

Colonies of typical *L. monocytogenes* as they appear when grown on *Listeria*-selective agar

The Anton test is used in the identification of *L. monocytogenes*; instillation of a culture into the conjunctival sac of a rabbit or guinea pig causes severe keratoconjunctivitis within 24 hours.

Listeria species grow on media such as Mueller-Hinton agar. Identification is enhanced if the primary cultures are done on agar containing sheep blood, because the characteristic small zone of hemolysis can be observed around and under colonies. Isolation can be enhanced if the tissue is kept at 4 °C for some days before inoculation into bacteriologic media. The organism is a facultative anaerobe and is catalase-positive and motile. *Listeria* produces acid, but not gas, in a variety of carbohydrates. The motility at room temperature and hemolysin production are primary findings that help differentiate listeria from coryneform bacteria.

The methods for analysis of food are complex and time-consuming. The present U.S. FDA method, revised in September 1990, requires 24 and 48 hours of enrichment, followed by a variety of other tests. Total time to identification takes five to seven days, but the announcement of specific nonradiolabelled DNA probes should soon allow a simpler and faster confirmation of suspect isolates.

Recombinant DNA technology may even permit two- to three-day positive analysis in the future. Currently, the FDA is collaborating in adapting its methodology to quantitate very low numbers of the organisms in foods.

Treatment

When listeric meningitis occurs, the overall mortality may reach 70%, from septicemia 50%, and from perinatal/neonatal infections greater than 80%. In infections during pregnancy, the mother usually survives. Reports of successful treatment with parenteral penicillin or ampicillin exist. Trimethoprim-sulfamethoxazole has been shown effective in patients allergic to penicillin.

A bacteriophage, *Listeria phage P100*, has been proposed as food additive to control *L. monocytogenes*. Bacteriophage treatments have been developed by several companies. EBI Food Safety and Intralytix both have products suitable for treatment of the bacterium. The U.S. Food and Drug Administration (FDA) approved a cocktail of six bacteriophages from Intralytix, and a one-type phage product from EBI Food Safety designed to kill *L. monocytogenes*. Uses would potentially include spraying it on fruits and ready-to-eat meat such as sliced ham and turkey.

Use as a Transfection Vector

Because *L. monocytogenes* is an intracellular bacterium, some studies have used this bacterium as a vector to deliver genes *in vitro*. Current transfection efficiency remains poor. One example of the successful use of *L. monocytogenes* in *in vitro* transfer technologies is in the delivery of gene therapies for cystic fibrosis cases.

Cancer Treatment

Listeria monocytogenes is being investigated as a cancer immunotherapy for several types of cancer.

A live attenuated *Listeria monocytogenes* cancer vaccine, ADXS11-001, is under development as a possible treatment for cervical carcinoma.

Epidemiology

Researchers have found *Listeria monocytogenes* in at least 37 mammalian species, both domesticated and feral, as well as in at least 17 species of birds and possibly in some species of fish and shellfish. Laboratories can isolate *Listeria monocytogenes* from soil, silage, and other environmental sources. *Listeria monocytogenes* is quite hardy and resists the deleterious effects of freezing, drying, and heat remarkably well for a bacterium that does not form spores. Most *Listeria monocytogenes* strains are pathogenic to some degree.

Routes of Infection

Listeria monocytogenes has been associated with such foods as raw milk, pasteurized fluid milk, cheeses (particularly soft-ripened varieties), ice cream, raw vegetables, fermented raw-meat sausages, raw and cooked poultry, raw meats (of all types), and raw and smoked fish. Most bacteria can survive near freezing temperatures, but cannot absorb nutrients, grow or replicate. *L. monocytogenes* ability to grow at temperatures as low as 0 °C permits exponential multiplication in refrigerated foods. At refrigeration temperature, such as 4 °C, the amount of ferric iron can affect the growth of *L. monocytogenes*.

Infectious Cycle

The primary site of infection is the intestinal epithelium, where the bacteria invade nonphagocytic cells via the "zipper" mechanism. Uptake is stimulated by the binding of listerial internalins (Inl) to E-cadherin, a host cell adhesion factor, or Met (c-Met), hepatocyte growth factor. This binding activates certain Rho-GTPases, which subsequently bind and stabilize Wiskott Aldrich syndrome protein (WAsp). WAsp can then bind the Arp2/3 complex and serve as an actin nucleation point. Subsequent actin polymerization creates a "phagocytic cup", an actin-based structure normally formed around foreign materials by phagocytes prior to endocytosis. The net effect of internalin binding is to exploit the junction-forming apparatus of the host into internalizing the bacterium. *L. monocytogenes* can also invade phagocytic cells (e.g., macrophages), but requires only internalins for invasion of nonphagocytic cells.

Following internalization, the bacterium must escape from the vacuole/phagosome before fusion with a lysosome can occur. Three main virulence factors that allow the bacterium to escape are listeriolysin O (LLO-encoded by *hly*) phospholipase A (encoded by *plcA*) and phospholipase B (*plcB*). Secretion of LLO and PlcA disrupts the vacuolar membrane and allows the bacterium to escape into the cytoplasm, where it may proliferate.

Once in the cytoplasm, *L. monocytogenes* exploits host actin for the second time. ActA proteins associated with the old bacterial cell pole (being a bacillus, *L. monocytogenes* septates in the middle of the cell, thus has one new pole and one old pole) are capable of binding the Arp2/3 complex, thereby inducing actin nucleation at a specific area of the bacterial cell surface. Actin polymerization then propels the bacterium unidirectionally into the host cell membrane. The protrusion formed may then be internalized by a neighboring cell, forming a double-membrane vacuole from which the bacterium must escape using LLO and PlcB. This mode of direct cell-to-cell spread involves a cellular mechanism known as paracytophagy.

Escherichia Coli

Escherichia coli (E. coli) bacteria normally live in the intestines of healthy people and animals. Most varieties of E. coli are harmless or cause relatively brief diarrhea. But a few particularly nasty strains, such as E. coli O157:H7, can cause severe abdominal cramps, bloody diarrhea and vomiting.

You may be exposed to E. coli from contaminated water or food — especially raw vegetables and undercooked ground beef. Healthy adults usually recover from infection with E. coli O157:H7 within a week, but young children and older adults have a greater risk of developing a life-threatening form of kidney failure called hemolytic uremic syndrome.

Symptoms

Signs and symptoms of E. coli O157:H7 infection typically begin three or four days after exposure to the bacteria, though you may become ill as soon as one day after to more than a week later. Signs and symptoms include:

- Diarrhea, which may range from mild and watery to severe and bloody
- Abdominal cramping, pain or tenderness
- Nausea and vomiting, in some people

Causes

Among the many strains of E. coli, only a few trigger diarrhea. One group of E. coli — which includes O157:H7 — produces a powerful toxin that damages the lining of the small intestine, which can cause bloody diarrhea. You develop an E. coli infection when you ingest this strain of bacteria.

Unlike many other disease-causing bacteria, E. coli can cause an infection even if you ingest only small amounts. Because of this, you can be sickened by E. coli from eating a slightly undercooked hamburger or from swallowing a mouthful of contaminated pool water.

Potential sources of exposure include contaminated food or water and person-to-person contact.

Contaminated Food

The most common way to acquire an E. coli infection is by eating contaminated food, such as:

- Ground beef. When cattle are slaughtered and processed, E. coli bacteria in their intestines can get on the meat. Ground beef combines meat from many different animals, increasing the risk of contamination.

- Unpasteurized milk. E. coli bacteria on a cow's udder or on milking equipment can get into raw milk.
- Fresh produce. Runoff from cattle farms can contaminate fields where fresh produce is grown. Certain vegetables, such as spinach and lettuce, are particularly vulnerable to this type of contamination.

Contaminated Water

Human and animal feces may pollute ground and surface water, including streams, rivers, lakes and water used to irrigate crops. Although public water systems use chlorine, ultraviolet light or ozone to kill E. coli, some outbreaks have been linked to contaminated municipal water supplies.

Private wells are a greater cause for concern because they don't often have any disinfecting system. Rural water supplies are the most likely to be contaminated. Some people also have been infected after swimming in pools or lakes contaminated with feces.

Personal Contact

E. coli bacteria can easily travel from person to person, especially when infected adults and children don't wash their hands properly. Family members of young children with E. coli infection are especially likely to acquire it themselves. Outbreaks have also occurred among children visiting petting zoos and in animal barns at county fairs.

Risk Factors

E. coli can affect anyone who is exposed to the bacteria. But some people are more likely to develop problems than are others. Risk factors include:

- Age. Young children and older adults are at higher risk of experiencing illness caused by E. coli and more-serious complications from the infection.
- Weakened immune systems. People who have weakened immune systems — from AIDS or drugs to treat cancer or prevent the rejection of organ transplants — are more likely to become ill from ingesting E. coli.
- Eating certain types of food. Riskier foods include undercooked hamburger; unpasteurized milk, apple juice or cider; and soft cheeses made from raw milk.
- Time of year. Though it's not clear why, the majority of E. coli infections in the U.S. occur from June through September.
- Decreased stomach acid levels. Stomach acid offers some protection against

E. coli. If you take medications to reduce your levels of stomach acid, such as esomeprazole (Nexium), pantoprazole (Protonix), lansoprazole (Prevacid) and omeprazole (Prilosec), you may increase your risk of an E. coli infection.

Complications

Most healthy adults recover from E. coli illness within a week. Some people — particularly young children and older adults — may develop a life-threatening form of kidney failure called hemolytic uremic syndrome.

Prevention

No vaccine or medication can protect you from E. coli-based illness, though researchers are investigating potential vaccines. To reduce your chance of being exposed to E. coli, avoid risky foods and watch out for cross-contamination.

Risky foods

- Cook hamburgers until they're 160 F. Hamburgers should be well-done, with no pink showing anywhere in the meat. But color isn't a reliable indicator of whether or not meat is done cooking. Meat — especially if grilled — can brown before it's completely cooked.

That's why it's important to use a meat thermometer to ensure that meat is heated to at least 160 F (71 C) at its thickest point.

- Drink pasteurized milk, juice and cider. Any boxed or bottled juice kept at room temperature is likely to be pasteurized, even if the label doesn't say so.
- Wash raw produce thoroughly. Washing produce won't necessarily get rid of all E. coli — especially in leafy greens, which provide many spots for the bacteria to attach themselves to. Careful rinsing can remove dirt and reduce the amount of bacteria that may be clinging to the produce.

Avoid cross-contamination

- Wash utensils: Use hot soapy water on knives, countertops and cutting boards before and after they come into contact with fresh produce or raw meat.
- Keep raw foods separate: This includes using separate cutting boards for raw meat and foods, such as vegetables and fruits. Never put cooked hamburgers on the same plate you used for raw patties.
- Wash your hands: Wash your hands after preparing or eating food, using the toilet, or changing diapers. Make sure that children also wash their hands before eating, after using the bathroom and after contact with animals.

References

- Ryan, Kenneth J.; Ray, C. George; Ahmad, Nafees; Drew, W. Lawrence; Lagunoff, Michael; Pottinger, Paul; Reller, L. Barth; Sterling, Charles R. (2014). "Pathogenesis of Bacterial Infections". Sherris Medical Microbiology (6th ed.). New York: McGraw Hill Education. pp. 391–406. ISBN 978-0-07-181826-1
- McFall-Ngai, Margaret (2007-01-11). "Adaptive Immunity: Care for the community". Nature. 445 (7124): 153–153. doi:10.1038/445153a. ISSN 0028-0836
- "Adult Health Advisor 2005.4: Bacteria in Urine, No Symptoms (Asymptomatic Bacteriuria)". Archived from the original on 2007-07-12. Retrieved 2007-08-25
- Epps SV, Harvey RB, Hume ME, Phillips TD, Anderson RC, Nisbet DJ (2013). "Foodborne Campylobacter: infections, metabolism, pathogenesis and reservoirs". International Journal of Environmental Research and Public Health. 10 (12): 6292–304. doi:10.3390/ijerph10126292. PMC 3881114. PMID 24287853
- Greenwood, David; Barer, Mike; Slack, Richard; Irving, Will (2012). "Bacterial Pathogenicity". Medical Microbiology, a Guide to Microbial Infections: Pathogenesis, Immunity, Laboratory Investigation, and Control (18th ed.). Edinburgh: Churchill Livingstone. pp. 156–167. ISBN 9780702040894
- Fujimoto, S; Amako, K (1990). "Guillain–Barré syndrome and Campylobacter jejuni infection". Lancet. 335: 1350. doi:10.1016/0140-6736(90)91234-2
- Wallis, M. R. (1994-03-01). "The pathogenesis of Campylobacter jejuni". British Journal of Biomedical Science. 51 (1): 57–64. ISSN 0967-4845. PMID 7841837
- Mikkola, Raimo. Food and Indoor Air Isolated Bacillus Non-Protein Toxins: Structures, Physico-Chemical Properties and Mechanisms of Effects on Eukaryotic Cells (PDF). p. 12. Retrieved 2015-10-24
- Gillespie SH, Hawkey PM (2006). Principles and practice of clinical bacteriology (2nd ed.). Hoboken, NJ: John Wiley & Sons. ISBN 9780470017968

Permissions

All chapters in this book are published with permission under the Creative Commons Attribution Share Alike License or equivalent. Every chapter published in this book has been scrutinized by our experts. Their significance has been extensively debated. The topics covered herein carry significant information for a comprehensive understanding. They may even be implemented as practical applications or may be referred to as a beginning point for further studies.

We would like to thank the editorial team for lending their expertise to make the book truly unique. They have played a crucial role in the development of this book. Without their invaluable contributions this book wouldn't have been possible. They have made vital efforts to compile up to date information on the varied aspects of this subject to make this book a valuable addition to the collection of many professionals and students.

This book was conceptualized with the vision of imparting up-to-date and integrated information in this field. To ensure the same, a matchless editorial board was set up. Every individual on the board went through rigorous rounds of assessment to prove their worth. After which they invested a large part of their time researching and compiling the most relevant data for our readers.

The editorial board has been involved in producing this book since its inception. They have spent rigorous hours researching and exploring the diverse topics which have resulted in the successful publishing of this book. They have passed on their knowledge of decades through this book. To expedite this challenging task, the publisher supported the team at every step. A small team of assistant editors was also appointed to further simplify the editing procedure and attain best results for the readers.

Apart from the editorial board, the designing team has also invested a significant amount of their time in understanding the subject and creating the most relevant covers. They scrutinized every image to scout for the most suitable representation of the subject and create an appropriate cover for the book.

The publishing team has been an ardent support to the editorial, designing and production team. Their endless efforts to recruit the best for this project, has resulted in the accomplishment of this book. They are a veteran in the field of academics and their pool of knowledge is as vast as their experience in printing. Their expertise and guidance has proved useful at every step. Their uncompromising quality standards have made this book an exceptional effort. Their encouragement from time to time has been an inspiration for everyone.

The publisher and the editorial board hope that this book will prove to be a valuable piece of knowledge for students, practitioners and scholars across the globe.

Index